THOMSON
DELMAR LEARNING

Fire Prevention: Inspection and Code Enforcement, 2nd Edition
by
David Diamantes

Executive Director:
Alar Elken

Executive Editor:
Sandy Clark

Acquisitions Editor:
Mark Huth

Development:
Dawn Daugherty

Executive Marketing Manager:
Maura Theriault

Channel Manager:
Fair Huntoon

Marketing Coordinator:
Brian McGrath

Executive Production Manager:
Mary Ellen Black

Production Editor:
Ruth Fisher

Printed in the United States of America
3 4 5 XX 05 04 03

For more information contact Delmar, 3 Columbia Circle, PO Box 15015, Albany, NY 12212-5015.

Or find us on the World Wide Web at http://www.delmar.com

Library of Congress Cataloging-in-Publication Data:

Diamantes, David.
Fire prevention : inspection and code enforcement / David Diamantes.—2nd ed.
p. cm.
Includes bibliographical references and index.
ISBN 0-7668-5285-7 (core)—
ISBN 0-7668-5286-5 (IG)
1. Fire prevention—Inspection—United States. I. Title.

TH9176 .D53 2002
363.37'7'0973—dc21

2002028771

NOTICE TO THE READER

Publisher does not warrant or guarantee any of the products described herein or perform any independent analysis in connection with any of the product information contained herein. Publisher does not assume, and expressly disclaims, any obligation to obtain and include information other than that provided to it by the manufacturer.

The reader is expressly warned to consider and adopt all safety precautions that might be indicated by the activities herein and to avoid all potential hazards. By following the instructions contained herein, the reader willingly assumes all risks in connection with such instructions.

The publisher makes no representation or warranties of any kind, including but not limited to, the warranties of fitness for particular purpose or merchantability, nor are any such representations implied with respect to the material set forth herein, and the publisher takes no responsibility with respect to such material. The publisher shall not be liable for any special, consequential, or exemplary damages resulting, in whole or part, from the readers' use of, or reliance upon, this material.

FIRE PREVENTION: INSPECTION AND CODE ENFORCEMENT

Second Edition

David Diamantes

THOMSON
DELMAR LEARNING

Australia Canada Mexico Singapore Spain United Kingdom United States

To my wife, Bonita, without
whose support and encouragement
this work would not have been possible.

Contents

Preface

WHY FIRE PREVENTION?

The incidence of fire in the United States says a lot about us as a people. Study after study has identified the loss from fire in our country as a national disgrace, yet the American people, the media, and all levels of government are unaware of or unable to recognize that a problem even exists. Many in the fire service are loath to dedicate staff and resources to the very effort that led to the development of the municipal fire service in this country: fire prevention. Many in fire service management view fire prevention as a political liability. Some have even gladly relegated the function to another agency or, worse, abandoned the effort in favor of other functions. The image of a fire inspector armed with a code book and a clipboard rarely makes the news unless it involves a politically sensitive issue.

The roots of today's fire service are anchored in the ashes of the great conflagrations of the late nineteenth and early twentieth centuries when fire brigades and salvage companies were commissioned by the stock fire insurance companies for the protection of insured properties. Perhaps we have forgotten where we came from. The protection of life and the preservation of property through education, regulation, and enforcement were and still are the job of the fire service. The very existence of many communities revolves around a few select industries. Most of the citizenry is employed within those industries, by their suppliers, or within the service sector that provides housing and basic goods and services.

When fire destroyed Farmland Foods' processing center in Albert Lea, Minnesota, on July 9, 2001, seven-hundred of Albert Lea's citizens were thrown out of work at the city's second largest employer.[1] The unemployment rate literally doubled overnight. The effect on the region's economy will be severe at best—devastating at worst. Mortgage and car loans will default; stores and restaurants will suffer a decline in business as belts are tightened. The economic trickle-down impacting the lives of literally thousands of people can be traced back to sparks from a welder's torch that ignited combustible materials. One of the most common and easily prevented causes of industrial fires has again changed the future of an entire region.

If a method to prevent earthquakes, tornadoes, or hurricanes were discovered, the public would demand swift implementation. Government and industry would devote millions of dollars and dedicate the best and the brightest to the project. Compare the pipe dream of controlling the weather with the reality of America's fire record, and the results are sobering:

> The fire problem in the United States, on a per capita basis, is one of the worst in the industrial world. Thousands of Americans die each year in fires, tens of thousands of peo-

ple are injured, and property losses reach billions of dollars. To put this in context, the annual losses from floods, hurricanes, tornadoes, earthquakes, and other natural disasters combined in the United States average just a fraction of the losses from fires. The public in general, the media, and local governments, however, are generally unaware of the magnitude and seriousness of fire to the communities and to the country.[2]

Fire in the United States, Eleventh Edition, August 1999

The frequency and severity of fires in America do not result from a lack of knowledge of the causes, means of prevention or methods of suppression. We have a fire "problem" because our nation has failed to adequately apply and fund known loss reduction strategies.[3]

Recommissioned Panel for America Burning, May 2000

America today has the highest fire losses in terms of both frequency and total losses of any modern technological society. Losses from fire at the high rate experienced in America are avoidable and should be as unacceptable as losses caused by drunk driving or deaths of children accidentally killed playing with guns.[4]

Recommissioned Panel for America Burning, May 2000

Too frequently an exhaustive report on (fire safety) conditions is treated by the municipal authorities with an indifference akin to contempt.[5]

National Association of Credit Men, Committee on Fire Insurance, June 1909

When I began my career as a firefighter, it never occurred to me that fires were not only a threat to my community, but a threat to my livelihood as well. I reasoned that the higher the fire frequency and the greater the damage, the more the public would realize just how vital my services were to the community. I could not have been more wrong. Every fire affects every member of the community through increased costs for government services, loss of jobs and productivity, increased insurance premiums, and erosion of the tax base. The municipal budget that funds the fire department is the same one that funds every other local government service and is entirely dependent on the local tax base. When forced to choose between government services, the public chooses education first. Who can resist the mantra "It's for the children"? Parks, libraries, public safety, and other services vie for whatever is left. Fire is always something that happens to *other people*. When forced to choose between a service that they feel directly affects them and those that only affect *other people*, which will the public choose?

The process of inspection and the enforcement of the fire prevention codes is one part (albeit an important one) in the fight against one of our nation's most significant problems. You have the means to play a part in an endeavor with a greater potential economic impact than controlling the weather. If you do not believe in preventing fires for the good of your community, believe it for your own job security.

HOW TO USE THIS BOOK

Throughout this text, the model codes and referenced standards are described as the tools of the trade for the fire inspector. Learning to be an inspector is, in fact, mostly a lesson in how to use these tools. Using this book without access to a model building and fire code is a bit like studying a how-to manual on fly-fishing without having a fly rod. The greatest challenge in writing the original text and in updating this second edition was attempting to describe the basic principles of the fire inspection process within the different code systems in use today. Much of the first chapter, on code administration, is devoted to describing the different approaches taken by local, state, and the federal government in adopting building and fire prevention regulations.

You might say that the model code system in our country is "in a state of flux." Two major camps have emerged from the initial efforts to develop a single set of construction and fire prevention codes for the United States. The International Code Council, made up of the Building Officials and Code Administrators International (BOCA), International Congress of Building Officials (ICBO), and Southern Building Code Congress (SBCCI) are in one camp, and the National Fire Protection Association (NFPA), Western Fire Chiefs Association (WFCA), and International Association of Plumbing and Mechanical Official (IAPMO) are in the other.

Both groups have developed or are in the process of developing a complete set of complementary codes for construction, fire prevention, and property maintenance. Why the two camps have chosen to take separate paths toward the same goal, which system of codes will provide the most balanced level of public safety versus cost, whose process of development is fairer or better, and who best represents the interests of the fire service belongs in a different book—that I doubt few would have the stamina or stomach to write.

In writing the first edition and in revising this second edition, I have attempted to describe the inspections process as broadly (across the various code groups) as possible. This edition is based on the 1999 editions of BOCA's *National Building Code* and *Fire Prevention Code*; ICBO's 1997 *Uniform Building Code* and *Fire Code*; SBCCI's 1999 *Standard Building Code* and *Fire Prevention Code*; NFPA 1, *Fire Prevention Code*; NFPA 5000, *Building Code*, second draft; and the Western Fire Chief's 2000 *Uniform Fire Code*, which is no longer published by ICBO.

As an inspector, you probably will not have a large say in the code system adopted within your jurisdiction anymore than you get to pick the mayor or city council. You will work with the tools they give you and attempt to provide the best service to the public and the business community you can. In the aftermath of a fire at the Imperial Foods processing plant in Hamlet, North Carolina, in 1991, in which twenty-five were killed and fifty-four injured, the local fire chief stated that the entire incident centered around one problem—lack of enforcement of existing codes.[6] *What* code is not as important as consistent, competent, and evenhanded inspection and enforcement.

NOTE TO INSTRUCTORS

In addition to the companion Instructor's Guide available from Delmar, PowerPoint slides, instructor handouts, and other useful instructional materials are available at www.efirecode.com.

Acknowledgments

The author and Delmar Learning gratefully acknowledge the comments and suggestions of the members of the review panel who contributed to this text. Their efforts have proved invaluable to the success of this text. The members of the review panel are:

Ed Smith
Crafton Hills College
Yucaipa, CA

Gail Ownby-Hughes
Chattanooga State Technical Community College
Chattanooga, TN

Llyod Stanley
Guilford Technical Commmunity College
Jamestown, NC

Timothy Flannery
John Jay College of Criminal Justice
New York, NY

Tommy Abercrombie
Tarrant County College
Fort Worth, TX

1

FIRE PREVENTION THROUGH REGULATION

Fire prevention inspection is no longer the exclusive bailiwick of uniformed firefighters assigned to fire prevention bureaus. Today's inspector is likely to be a civilian combination inspector who performs building inspections, property maintenance inspections, and fire inspections, all under the authority of the fire chief. Every conceivable combination of uniform/civilian staffing, enforcement agency organization, and division of responsibility has been tried or is in use today in the United States. Tight budgets and the philosophy of making government leaner and smarter drive the efforts.

A few fire prevention offices have the sole responsibility of enforcing the fire prevention code. Others investigate fires and have law enforcement responsibilities. Still others are involved in plan review, acceptance testing, and the inspection of new construction. The arrangement with the most potential significant impact is the office that combines the traditional responsibilities of the building department, fire inspection office, and fire investigation office under one director within the fire department.

The benefits of a coordinated effort include increased public safety and economic attractiveness for the jurisdiction. When fire department operational effectiveness and firefighter safety are considered from the planning stage through development, with constant feedback through the investigation process, everyone wins including the business community.

With the mix of roles and responsibilities within many jurisdictions, legal authority can sometimes become a confusing issue. Many of us wear many hats, such as fire marshal, fire inspector, fire investigator, building inspector, or a combination of some or all of the above. These roles may exist in statute or regulation, and may overlap. *Always remember which hat you have on: You have only the authority and responsibility of your position.*

Code Administration

Learning Objectives

Upon completion of this chapter, you should be able to:

- Describe the model code process and name the model code organizations.
- Describe the origin of the first *National Building Code*.
- Describe the code adoption process used by state and local governments and explain the basis for their authority to enact such regulations.
- Describe the economic forces behind the development of a single national model code.

Essential qualifications for the successful Inspector are ability to recognize fire hazards when he sees them, a logical and practical understanding of how to safeguard them, persistence in securing necessary corrections, and absolutely impartial and accurate observation and reporting of conditions found.[1]

(From *Field Practice—An Inspection Manual for Property Owners, Fire Departments and Inspection Offices*, 1922, National Fire Protection Association.)

Severus and Caracalla, Emperors, to Junian Refianas, Brigadier General of the Vigiles: Greetings! You are hereby authorized to punish with the rod or cat-of nine-tails, the janitor, or any of the inhabitants of a house in which fire has broken out through negligence. In case the fire should have been occasioned not by negligence but by crime, you must hand the incendiaries to our friend, Fabius Septimiamus Cilo, Prefect of the City.[2]

(Rome, 1st century A.D. Taken from an ancient inscription discovered in the headquarters of the Vigiles or Roman fire brigade, current site of Fire and Police Department Headquarters, Rome, Italy.)

code
a systematically arranged body of law or rules; *when and where to do or not to do something*

Fire prevention as practiced by the Romans in the first century was quite different from the system of technical **codes** that has evolved and is used throughout most of the industrialized world. The "code" enforced by the Vigiles, a seven-thousand-man fire department, was short and sweet: *Have a fire? Get punished.* The first-ever building regulations, decreed by King Hammurabi, founder of the Babylonian empire, were equally succinct. *If the building collapses and kills the owner, the builder shall be put to death. If the owner's son is killed, the builder's son shall be put to death.*

NOTE
Early societies regulated behavior by ensuring punishment in the event of failures.

These early societies regulated behavior by ensuring punishment in the event of failures. They specified the failure to be avoided and the punishment to be meted out, but they did not specify the procedures to be followed in order to prevent the failure. Technical codes, which specified construction methods and materials, requirements for material handling, and regulations for certain processes, came much later, but not until after centuries of conflagrations and at a cost of countless lives and dollars.

model code
a code generally developed through the consensus process through the use of technical committees developed by an organization for adoption by governments; e.g., the Uniform, Standard, and BOCA/National codes

adopt
to formally accept and put into effect

The history of fire in America dates back to the time that our forefathers first stepped off of the boat. The first permanent colony in what was to become the United States was established by the English in Jamestown, Virginia, in 1607. In 1608 fire destroyed most of the buildings and provisions.[3] Disastrous fires in Boston in 1676 and 1679 led to the establishment of regulations that required all buildings to be of noncombustible construction (stone or brick and covered with slate or "tyle"). Unfortunately, Boston's regulations were never enforced, and the fires continued.

In 1871, Chicago burned for almost two days killing 250 people, destroying 17,000 buildings, and leaving almost 100,000 homeless. Five years later, the city enacted building and fire prevention codes. A series of fires around 1900, including a waterfront fire in Hoboken, New Jersey, where 300 were killed and a fire in Baltimore that resulted in over $50 million dollars in direct damage and the unemployment of 50,000 people, finally prompted the insurance industry to act.[4]

MODEL CODES

In 1905 The National Board of Fire Underwriters (NBFU), which later became the American Insurance Association or AIA, published the first *National Building Code*.[5] Unlike today's **model codes**, the *National Building Code* was developed by staff of the NBFU, not through the use of technical committees and the consensus process. The *National Building Code* was published by the NBFU (later to become

■ NOTE
Unlike today's model codes, the *National Building Code* was developed by staff of the NBFU, not through the use of technical committees and the consensus process.

■ NOTE
The *National Building Code* was made available for state and local governments to adopt at no charge.

the AIA) through 14 editions over 75 years. The *National Building Code* pioneered the very fundamental principles that drive our buildings codes today, such as building height and area restrictions, fire-resistive construction, shaft enclosure, fire-walls and doors, firestopping, means of egress, and fire suppression systems.[6]

What was most significant about NFBU's action was that the *National Building Code* was made available for state and local governments to **adopt** at no charge. Governments were given their first opportunity to adopt a building code developed by experts from outside their political jurisdictions, free from local political influences.

The NBFU also developed a fire prevention code that was intended to be a companion document. The *National Building Code* was discontinued by AIA after 1976.[7] The right to publish the *National Building Code* was later acquired by Building Officials and Code Administrators International (BOCA), and in 1984 the *BOCA Basic Building Code* became the *BOCA Basic/National Building Code*. In 1987 the word *Basic* was dropped altogether leaving the *BOCA National Building Code (BNBC)*.

MODEL CODE ORGANIZATIONS

The International Code Council

On December 9, 1994, the International Code Council (ICC) was officially organized as a nonprofit, umbrella organization consisting of representatives of BOCA, International Conference of Building Officials (ICBO), and Southern Building Code Congress International (SBCCI). The sole purpose of the ICC was to facilitate the development and maintenance of the International Codes and to coordinate related supporting activities.[8]

The 1990s ushered in a series of political developments that brought our regional system of model code development under close scrutiny. The North American Free Trade Agreement (NAFTA) and actions with the European Common Market that effectively eliminated many barriers to European trade were cause for growing concern. Was our system of regional model codes with technical disparities between each a threat to American competitiveness? The response from the three model code groups was to form the ICC and begin development of the international codes. By 1996, the *International Plumbing, Mechanical*, and *Private Sewage Disposal Codes* were published in 2000. Under the agreement establishing the ICC, the members groups agreed to discontinue publishing their individual codes as of 2000.

ordinance
a law of an authorized subdivision of a state, such as a county, city, or town

Building Officials and Code Administrators International (BOCA)

BOCA was established in 1915 by building officials from nine states and Canada as the Building Officials Conference of America,[9] "to discuss the principles underlying **ordinances** related to building."[10] In 1950 BOCA first published the

Basic Building Code. The organization maintained building, mechanical, fire prevention, plumbing, and property maintenance codes through 1999. BOCA generally serves the northeastern and midwestern states.

International Conference of Building Officials (ICBO)

ICBO was established as the Pacific Building Officials Conference in 1921, and in 1927[11] published the *Uniform Building Code (UBC)*, and numerous related codes that addressed housing maintenance, building security, outdoor signs, and building conservation. The *Uniform Fire Code* was maintained by the Western Fire Chief's Association and ICBO through the 1988 edition. The International Fire Code Institute (IFCI) was formed and published the codes since 1991. The *Uniform Mechanical Code* is maintained by the International Association of Plumbing and Mechanical Officials (IAPMO). ICBO serves the western and midwestern states.

Southern Building Code Congress International (SBCCI)

SBCCI was established in 1940 and in 1945 first published the *Standard Building Code (SBC)*.[12] In cooperation with the Southeastern Association of Fire Chiefs, SBCCI published the *Standard Fire Prevention Code* through 1999. Like BOCA and ICBO, SBCCI developed and maintained numerous related codes that addressed housing maintenance, building security, outdoor signs, and building conservation. SBCCI serves the southeast.

National Fire Protection Association (NFPA)

standard
a rule for measuring or a model to be followed; *how to do something*, what materials to use (also known as *referenced standard*)

NFPA was organized in 1896 and incorporated in 1930. The existence of five distinct electrical codes in the United States as well as nine radically different **standards** for sprinkler pipe size and spacing within 100 miles of Boston[13] were instrumental in steering fire underwriters into forming an association to work toward uniformity in standards for fire protection. The group was to be known as the National Fire Protection Association.

By 1904 NFPA's active membership included 38 stock fire insurance boards and 417 individuals, most of whom were from the insurance industry. The first fire department officer to join NFPA was Battalion Chief W. T. Beggin of the FDNY who became a member in 1905. H. D. Davis, state fire marshal of Ohio joined the same year. In 1911 NFPA began maintaining the *National Electrical Code* and has published it ever since.[14]

NFPA produces almost 300 codes, standards, and recommended practices developed by more than 205 technical committees. Among these, NFPA 70, *National Electrical Code*, in use throughout most of the United States, may well be the most widely used code of its type in the world. Work is currently underway on NFPA 5000, *Building Code*.

NFPA 101, *The Lifesafety Code*, addresses occupant safety in buildings with regard to the establishment and maintenance of exit facilities. It "does not attempt to address those general fire prevention or building construction features that are normally a function of fire prevention and building codes."[15] NFPA 101 separates buildings into thirteen occupancy types and then establishes requirements for new as well as existing buildings. The physical characteristics and limitations of the occupants as well as the structure are considered. This comprehensive standard addresses the building, the occupants, and the processes that occur. Many jurisdictions enforce the provisions of NFPA 101 in addition to their building and fire prevention codes. Often, certain provisions of NFPA 101 are more stringent and supersede those of the other codes.

THE MODEL CODE PROCESS

The ICC code change process is based on those used in the past by BOCA, ICBO, and SBCCI, with a significant change in the voting process. In previous years, voting was limited to active code officials representing member government jurisdictions. The number of votes was based on the jurisdiction's population. Representatives of industry were permitted to propose changes and testify at hearings, but were not permitted to vote in the final action portion of the process. This was the fundamental difference between the ICC and NFPA processes. The intent was to limit voting to persons without a financial interest in the process. Those who disagreed with the system claimed that without industry votes, the codes were not "true consensus codes." The counterargument stated that the system resulted in the "best codes that money can buy." Where the truth lies, only time will tell.

Step 1. Code changes are submitted by interested persons or groups. They are editorially reviewed by staff, published, and distributed to the membership.

Step 2. Technical committees hear testimony on the proposed changes at public hearings at the national conferences of the model code groups. Committees approve, deny, or approve a modified form of each code change.

Step 3. Immediately after the vote by the technical committee, the moderator asks for a motion from the floor to challenge or overturn the committee action.

Step 4. All members of BOCA, ICBO, and SBCCI in attendance at the public hearing are eligible to vote on floor motions.

Step 5. The results of the public hearings are published and distributed to interested parties. This action is designed to give attendees at the upcoming annual meeting an opportunity to consider specific objections to the results of the public hearings and prepare testimony for the public comment process that precedes the final action consideration, of Step 7.

Step 6. Discussion during the public comment process is limited to items that meet the following criteria: Items for which a public comment has been submitted; Items which may result in a technical inconsistency between the International Residential Code and the associated ICC International Code and Items which received a successful assembly action at the public hearing.

Step 7. During final action consideration hearings, testimony is limited to code change proposals that have an assembly action or public comment (from Step 5) and proposals on which the public hearing actions result in a technical inconsistency between the International Residential Code and the associated ICC International Code.

Step 8. The final action on all proposed code changes is published as soon as practicable after the determination of final action. Code changes are published with the next edition of the code.

NFPA's process differs. Their twelve-step process requires 104 weeks and relies on meetings of the 205 individual technical committees.

Step 1. NFPA calls for proposed changes to its documents.

Step 2. Committees meet, act on proposals, and also develop their own recommended changes.

Step 3. Committees vote by letter ballot. If two-thirds of the committee approve, the report continues in the process. Lacking two-thirds the report is returned to the committee.

Step 4. Reports are published for public review.

Step 5. Committees meet and act on each public comment.

Step 6. Committee votes by letter ballot on comments. Again, two-thirds majority is required for the report to proceed.

Step 7. Supplementary report is published for public review.

Step 8. NFPA membership meets and acts on committee reports.

Step 9. Committee votes on amendments approved at NFPA meeting.

Step 10. Twenty-day opportunity to file a complaint with the Standards Council by any person dissatisfied with the committee action.

Step 11. Standards Council acts on complaints.

Step 12. Twenty-day opportunity to appeal action of the Standards Council to the NFPA Board of Directors.

The members of NFPA's technical committees can be found on the opening pages of every NFPA standard. The 1995 edition of NFPA 251, *Standard Methods of Tests of Fire Endurance of Building Construction and Materials*, lists twenty-seven members and eleven alternates. My unscientific assessment based solely on the roster is that the committee is composed of two fire service representatives, six representatives of the insurance industry, six representatives of various testing and research labs, two representatives from major universities, and the remainder representing various industry groups from hotels to steel to textiles.[16]

Does any one group have control of the committee? NFPA has a thirteen-member Standards Council and has adopted *Regulations Governing Committee Projects* specifically to keep the standards-making process fair and above board. Each committee member is classified according to guidelines into one of nine categories:[17]

Manufacturer

User

Installer/Maintainer

Labor

Enforcing Authority

Insurance

Special Expert

Consumer

Applied Research/Testing Laboratory

Each committee is structured so that not more than one-third of the membership represents a single interest.

Is the fire service adequately represented in the process? Many in the fire service would emphatically answer no. Committee membership is an expensive proposition. Travel and lodging costs associated with two committee meetings a year are difficult for most fire departments to justify to skeptical city managers and budget offices. And sometimes unfortunately, such skepticism exists within our own ranks. Some chiefs see little advantage in participating in the process. Industry recognizes the cost of committee participation for what it really is, a legitimate business expense. Most local governments either lack the perspective or the resources.

International Fire Code Institute (IFCI)

A discussion of the model code process must include a special mention of the Western Fire Chiefs Association and its role in the development and maintenance of the *Uniform Fire Code (UFC)*. The *UFC* was initially developed by the California Fire Chiefs Association and published by ICBO and the Western Fire Chiefs Association in 1971. The International Fire Code Institute (IFCI) was created in 1991 to maintain the *UFC*. Through IFCI, the *UFC* was maintained through a process similar to that used by the ICC. In 2000 the UFC was published by WFCA alone.

Since 1988, the *Standard Fire Prevention Code (SFPC)* process has been administered by the Southeastern Association of Fire Chiefs under a similar process.

Unlike the IFCI process and the *Uniform Fire Code*, SBCCI remains the publisher and copyright holder of the code.

Council of American Building Officials (CABO)

CABO was established to promote uniformity among the model codes while the three model code groups retained their autonomy.[18] The first cooperative code development effort between BOCA, ICBO, and SBCCI was the CABO *One- and Two-Family Dwelling Code* in use in much of the country today, and the CABO *Model Energy Code*. The responsibility for both documents has been passed to the ICC by the directors of CABO.

AUTHORITY TO ENFORCE THE MODEL CODES

Rarely does a fire inspector ponder the source of his authority to inspect and order compliance as he drives between inspections. Most of an inspector's time between stops is filled with the more mundane world of hurried phone calls to the office, last minute inspection requests, and reports of inoperative fire protection systems and blocked exits. But the knowledge of the basis for that authority as well as the responsibilities that go with it should always be in the back of the inspector's mind.

■ NOTE
The legal basis for the promulgation and enforcement of building and fire prevention regulations by state and local governments actually originates within the Tenth Amendment to the United States Constitution.

The legal basis for the promulgation and enforcement of building and fire prevention regulations by state and local governments actually originates within the Tenth Amendment to the United States Constitution:

> The powers not delegated to the United States by the Constitution, nor prohibited by it to the states, are reserved to the states respectively, or to the people.

Police power, the fundamental power of the state to place restraints upon the personal freedom and property rights of individuals for the protection of the public health, safety, and welfare, is an inherent power of the states, possessed by them before adoption of the United States Constitution and reserved to them under the Tenth Amendment.[19]

Dillon's rule
legal ruling issued by Chief Justice John Forrest Dillon of the Iowa Supreme Court in the late 1800s whereby local governments possess only those powers expressly granted by charter or statute

Notice, however, that the Tenth Amendment made no mention of those powers reserved for municipalities. That is because there are none. The powers of local governments are only those powers delegated by the states. In the late 1800s, Chief Justice John Forrest Dillon of the Iowa Supreme Court wrote what was to become known as **Dillon's rule**. When arguments came before the Iowa Supreme Court regarding the powers of local governments, Justice Dillon's opinion became the rule for Iowa and later most of the states.

The late 1800s has been described as the low point for local government in our country. Political corruption was rampant. Lord Byron of England wrote at the time, "There is no denying that the government of cities is one of conspicuous failure of the United States."[20] Against this backdrop, Dillon's court heard arguments that these same governments had powers beyond those granted by the Constitution. Dillon's opinion in the case has been described as "absolutely the most important legal proposition in all of local governmental law."[21]

statute
a law enacted by a state or the federal legislature

Simply stated, under Dillon's rule, local governments have only those powers expressly conferred by the state constitution, state **statutes**, or home rule charter; those powers implied in or incidental to the powers expressly granted, and those powers essential to the declared purposes of the municipality.[22]

NOTE
Simply stated, under Dillon's rule, local governments have only those powers expressly conferred by the state constitution, state statutes, or home rule charter.

CODE ADOPTION

We will assume then that the laws of your state or your city charter have enabled the governing body to adopt regulations to protect the public from the threat of fire and explosions. Several states have adopted fire prevention regulations at the state level, leaving local government with only the decision of which agency will enforce them. Either way, in most of the United States, one of the three model fire prevention codes is adopted by the governing body. Hopefully, it is a companion document to the building code in use within the jurisdiction.

NOTE
The model code groups make their codes available to jurisdictions free of charge for adoption by reference only.

The adoption process is structured and is spelled out within the law. Generally, public notice of the jurisdiction's intent to adopt or amend the code is required. Publication of the time and date of public hearings and the availability of the proposed code text for public inspection prior to the hearings is required to appear in a newspaper serving the jurisdiction. The governing body acts on the proposed ordinance after hearing testimony from government officials and interested members of the public. The adopting ordinance also includes an effective date and designates the particular agency or agencies charged with the enforcement of the ordinance.

adoption by reference
method of code adoption in which the specific edition of a model code is referred to within the adopting ordinance

adoption by transcription
method of code adoption in which the entire text of the code is published within the adopting ordinance

Two methods of adoption are by **reference** and by **transcription**. Adoption by reference is the more common method, with the adopting ordinance referring to the specific edition of the model code to be enforced. Adoption by transcription requires the entire code to be published within the ordinances of the municipality. The model code groups make their codes available to jurisdictions free of charge for adoption by reference only. Publication of the documents as is necessary in adoption by transcription is only permitted under royalty-bearing licenses granted by the model code groups.

MINIMUM CODES AND MINI/MAXI CODES

In some states, local governments are only empowered to adopt regulations that exceed the requirements of minimum standards adopted at the state level. This system gives local government limited powers.

mini/max code
a code developed and adopted at the state level for either mandatory or optional enforcement by local governments, and which cannot be amended by the local governments

Under the **mini/maxi** concept, local governments are precluded from adopting any regulations. Uniform regulations are adopted at the state level and cannot be locally amended. Mini/maxi codes are generally favored by architects, builders, and contractors as being good for business by providing uniform regulations. Lobbying efforts can be targeted at the state capital and not spread across the numerous political subdivisions. Mini/maxi codes are generally opposed by fire officials and local governing bodies as an unnecessary intrusion into a local issue by a governing body that may be out of touch with the needs of the people.

Kentucky, Minnesota, Montana, New Jersey, North Carolina, Oregon, and Virginia are among the states with some form of mini/maxi code.[23] More will follow with our national push to reduce trade barriers. There is one positive note in the mini/maxi saga. It forces us in the fire service to get our heads out of the sand and participate in the code development process at the state and national levels.

NOTE
Under the mini/maxi concept, local governments are precluded from adopting any regulations; uniform regulations are adopted at the state level and cannot be locally amended.

LOCAL AMENDMENT OF THE MODEL CODES

Model codes are, by their very nature, designed to address the fire problems faced by most jurisdictions. This is one of the relative strengths of the entire process. *Most* of the problems in *most* jurisdictions are addressed by provisions based on our national and regional fire experience. What about the condition then that exists solely in one jurisdiction? Or perhaps the desire of a community to have standards of safety that far exceed the national norm?

NOTE
Model codes are, by their very nature, designed to address the fire problems faced by most jurisdictions.

These needs are addressed through amendments to the model codes that are made at the state or local level and enacted within the code adoption process. A city or county basically passes an ordinance stating "we hereby adopt the nineteen ninety-something edition of this model code with the following changes."

Many jurisdictions have adopted code requirements for sprinkler protection, fire department water supply, emergency vehicle access, and explosives or hazardous materials handling and storage that far exceed the model code provisions. Local amendments must be nondiscriminatory, technically valid, and justified by real need. Regardless how small any code provision may seem, there is a cost associated in some way. In the end, somebody is going to have to pay the bill.

Summary

The model code process is less than a century old in the United States and continues to evolve today. The three model building code groups—BOCA, ICBO, and SBCCI—joined forces in 1994, establishing the International Code Council (ICC). The ICC now publishes the international fire, building, mechanical, and plumbing codes. The "I" codes are distributed and maintained by BOCA, ICBO, and SBCCI. Work is underway to merge the three groups into a single organization under the umbrella of the ICC.

NFPA, traditionally the developer of the *National Electrical Code*®, the *Life Safety Code*, and many of the standards referenced by the ICC codes, is developing a building code, NFPA 5000, *Building Code*. Only time will prove the wisdom (or lack of it) of NFPA's foray into development of a building code.

The ICC and NFPA develop codes, standards, and recommended practices by consensus through the use of unpaid technical committees comprising experts in various fields. Model codes must be adopted on the state or local level in order to be enforced by the local enforcing agency. The adopting ordinance must specify the specific edition (year) that is to be adopted. The Department of Defense "adopts' model codes and standards as regulations through inclusion in their *Military Handbook for Fire Protection for Facilities Engineering, Design, and Construction*, MIL-HDBK-1008C.

Review Questions

1. Name four model code organizations.
   1. ______________________
   2. ______________________
   3. ______________________
   4. ______________________
2. Before a model code can be enforced within a jurisdiction it must be __________ by the governing body.
3. Changes to the model building and fire prevention codes are voted on by __________.
4. The fundamental power of the state to place restraints on personal freedoms and property rights is called __________.
5. Name two methods of code adoption
   1. ______________________
   2. ______________________
6. A uniform code that is adopted at the state level and that cannot be amended by local governments is called a ____________ code.
7. Which amendment to the United States Constitution is the basis for the promulgation and enforcement of building and fire prevention regulations by state and local governments?
   ______
8. What umbrella organization was organized by the model code groups to facilitate the

development of the international building and fire prevention codes? _______

9. What legal principle states that local governments have only those powers conferred by charter or state statute? _______

10. Model codes are available for local governments to adopt by __________ free of charge.

Discussion Questions

1. What organization developed the first National Building Code and why was the development significant?
2. What was a fundamental difference between the process used by the National Board of Fire Underwriters (NFBU) in the development of their codes and that used by the model code groups today?
3. Why are mini/max codes favored by many business interests?

Inspection

Learning Objectives

Upon completion of this chapter, you should be able to:

- Describe the limits of the *right of entry* provisions contained in the model fire prevention codes.
- Explain why warrants must be secured in order to inspect certain buildings and structures.
- Describe the permit and inspection models and the benefits of each.
- Describe the legal requirements for written notices and orders and how freedom of information laws impact the fire prevention bureau.
- Describe how the appeals process within the model codes is established and designed to operate.

The heart of any fire hazard law is INSPECTION, the value of which is generally underestimated as it is not spectacular or colorful.[1]

(*Virginia Advisory Legislative Council to the Governor, November 18, 1947*)

Just how effective is the inspections process at preventing the occurrence of fire or of minimizing the impact of hostile fires? Some would venture that comparison of any process or premises to "nationally recognized good practice" must result in the identification of deficiencies, and if corrections, repairs, or adjustments were made, things would simply have to be safer. That is probably true, but a more scientific approach was used in a study in the late 1970s that clearly identified the potential value of the inspection process.

In 1978, the National Fire Protection Association published the results of a federally funded study under the title *Fire Code Inspections and Fire Prevention: What Methods Lead to Success?* The study was cooperatively conducted by NFPA and the Urban Institute, with funding and support from the U.S. Fire Administration, the National Science Foundation, and twelve metropolitan fire departments from across the United States. The study was undertaken to determine whether some fire code inspection practices *actually resulted in fewer fires, lower fire loss, and fewer civilian casualties.* The study focused on properties covered by fire codes, excluding one- and two-family dwellings.

The study used fire data from seventeen cities and one metropolitan county, and determined that fires due to visible hazards than can be directly remedied by inspectors were responsible for 4–8% of all fires. Fires due to carelessness or foolish actions or electrical or mechanical failures (sometimes due to lack of maintenance) accounted for 40–60% of all fires. The remaining fires (32–66%) were identified as incendiary, suspicious, or natural cause fires, and were considered unpreventable by inspection.[2] When this information was compared to inspection practices, the findings underscored the effectiveness of a routine fire inspection program.

Fire rates appeared to be significantly lower in cities that annually inspected all, or nearly all, public buildings. Jurisdictions that did *not* annually inspect most public buildings had rates of fires exceeding $5,000, which were more than *twice* that of those that *did* inspect them.

Cities that used fire-suppression companies for a large share of their routine fire code inspections had substantially lower rates of fire, presumably because cities that exclusively used fire prevention bureau inspectors did not have sufficient personnel within the bureau to make annual inspections at all public buildings.

Fires due to carelessness or foolish actions or electrical or mechanical failures (40–60% of all fires) were not preventable through direct action by inspectors, because these causes are not readily visible during an inspection. Yet the rate at which these fires occurred was significantly lower in cities with regular inspection programs, indicating that the department that inspects more frequently has more opportunities to motivate occupants. This statistic also underscores the value of company inspections, recognizing the fact that although company inspections cannot be as in-depth, due to the differences in training between inspectors in the bureau and those with other responsibilities, inspections by in-service companies have a significant impact on the incidence of fire within the jurisdiction.

The opportunity to preplan hazards, provide building familiarization, and the chance to establish cordial relations with the community can literally make in-service inspections the excuse that every chief wants to advertise the department. Training for company inspections should be prefaced with the fact that the goals of the in-service inspection program are building familiarization to increase firefighter safety, fire and hazard reduction, and community relations—*in that order.*

■ NOTE
The inspector's duties and responsibilities in code enforcement are outlined within the body of the code.

The inspection process is the very backbone of the fire prevention program. Having a fire prevention code in force within a jurisdiction is of little benefit to the public without an effective inspection and enforcement program. Unlike the Vigiles of the first century in Rome, the fire inspector's duties do not include punishing those guilty of violating the code or causing a fire through negligence. The inspector's duties and responsibilities in code enforcement are outlined within the body of the code.

■ NOTE
Up-to-date information regarding legal developments within the code enforcement field, periodic training in the subject, and competent legal advice are absolute necessities for the fire code official.

How the inspector performs those functions and his or her conduct in the performance of those duties are governed by various federal, state, and local laws. From local ordinances that prohibit conflict of interest for public officials, to state tort claims laws, to federal civil rights statutes, fire inspections personnel are subject to myriad laws in the day-to-day performance of their duties. Up-to-date information regarding legal developments within the code enforcement field, periodic training in the subject, and competent legal advice are absolute necessities for the fire code official. This text touches lightly on the legal aspects of code enforcement, concentrating instead on the technical aspects of fire code inspection and enforcement. The need for training in the legal aspects of code enforcement cannot be overemphasized.

LEGAL ASPECTS

The inspection of certain structures and premises is mandated by two of the model fire prevention codes:

> *The code official shall inspect all structures and premises, except single-family dwellings and dwelling units in two-family and multi family dwellings for the purposes of ascertaining and causing to be corrected any conditions liable to cause fire, contribute to the spread of fire, interfere with fire-fighting operations, endanger life or any violations of the provisions or intent of this code or any other ordinance affecting fire safety.* (F-108.1, *1999 BOCA National Fire Prevention Code*)
>
> *The fire prevention bureau shall inspect, as often as necessary, buildings and premises, including such other hazards or appliances designated by the chief for the purpose of ascertaining and causing to be corrected any conditions which would reasonably tend to cause fire or contribute to its spread, or any violation of the purpose or provisions of this code and any other law or standard affecting fire safety.* (103.3.1.1, *2000 Uniform Fire Code*)

Although the *Standard Fire Prevention Code* does not contain a specific duty to inspect clause, SFPC 101.2.1 (SFPC, 1994, p. 1) states *this code is hereby declared remedial*, or intended to ameliorate or make better, . . . *through safety to life and property from fire, explosion and other hazards incidental to the use and occupancy of buildings, structures, or premises.*

The 2000 *International Fire Code* and NFPA 1 *Fire Prevention Code* do not mandate inspections specifically. They *authorize* inspections in order to ensure compliance with the code, but do not contain specific duty to inspect clauses. Code sections that appear to mandate the inspection of all structures may clarify the intention of the code writers that fire prevention codes are applicable to any and all structures or premises, but they have not been found by the courts to impose liability on the jurisdiction for failure to carry out this mandate.

Under common law, local governments have not been held liable for injuries resulting from enforcement or *failure to enforce* statutes or ordinances, as was clearly underscored in *Grogan v. Commonwealth*, a suit arising out of the disastrous Beverly Hills Supper Club fire in 1977. The City of Southgate, Kentucky, and the State Fire Marshal's Office were named as defendants arising from their failure to enforce the state building code and that the failure resulted in 165 deaths. Both the city and state were dismissed from the suit during preliminary motions by the trial court. The Kentucky Supreme Court later upheld the dismissal.[3]

NOTE

A common misconception is that single family dwellings and the dwelling units in multiple-family dwellings are exempt from the provisions of the code.

NOTE

The model codes do not exclude dwellings or dwelling units from their scopes; dwelling units are only exempted from routine inspection.

Applicability of the Code

The fire prevention code applies to *all structures and premises* (F-102.1 *BNFPC*); *buildings and premises* (101.2 *UFC*); and "*every existing building or structure and any existing appurtenances*," (101.3.1 *SFPC*). A common misconception is that single-family dwellings and the dwelling units in multiple-family dwellings are exempt from the provisions of the code. This misconception is a serious misinterpretation of the code or of the Fourth Amendment that could lead to dire safety consequences for the public as well as potential liability for the fire official.

The model codes do not exclude dwellings or dwelling units from their scopes. Dwelling units are only exempted from routine inspection. Timothy Callahan, writing in *Fire Service and the Law*, describes the balance between the legitimate police power of the state and the public's right of privacy:

> Subject to issuance of appropriate notice, there is no doubt that the right to inspect a private dwelling could be considered a reasonable exercise of the police power. This is true where there is reason to believe that conditions exist which make the property a fire menace to the community. But where a residence is so isolated from other occupancies or properties that it would not be considered a hazard, even if on fire, the right of privacy of the dwelling's owners might be considered paramount to the police power.[4]

However, Supreme Court decisions such as *Mapp v. Ohio*, *Frank v. Maryland*, and *Camara v. Municipal Court* have clearly underscored the limits on police power

warrant
a legal writ issued by a judicial officer commanding an officer to arrest a person, seize property, or search a premises

■ NOTE
The nonconsensual warrantless entry of a building is an especially serious breach, unless the entry is in response to an emergency.

■ NOTE
The protection of the Fourth Amendment is not limited to dwellings.

imposed by the Fourth Amendment. The nonconsensual **warrantless** entry of a dwelling is an especially serious breach, unless the entry is in response to an emergency.

In *Michigan v. Tyler*, the Court held that a "burning building clearly presents an emergency of sufficient proportions to render warrantless entry reasonable. Indeed, it would defy reason to suppose that firemen must secure a warrant or consent before entering a structure to put out the blaze."[5]

The protection of the Fourth Amendment is not limited to dwellings. In *See v. City of Seattle*, the Court held that "the businessman, like the occupant of a residence, has the constitutional right to go about his business free from unreasonable entries upon his private commercial property."[6] The right of the businessman to deny government officials access to those areas of his premises that are not open to the public is protected by the Fourth Amendment. Unless permission to enter and conduct an inspection is obtained, a warrant must be secured by the inspector.

Right of Entry

Each of the model codes contains a right of entry clause that permits the fire official to enter all structures and premises at reasonable times to conduct inspections. The inspector's right of entry is limited however:

- The fire prevention code cannot authorize an unconstitutional inspection. The inspector must identify himself, present credentials, and seek permission to conduct the inspection. Permission to inspect must be obtained or an inspection warrant must be secured prior to conducting the inspection, unless the inspector has reason to believe that a life-threatening condition exists.
- Inspections must be aimed at securing or determining compliance with the code, not with gathering information for other purposes such as the enforcement of other laws or ordinances.[7]
- Permission to conduct inspections must be requested during reasonable hours, normally the hours that the business is open or staff is present.
- Inspection of those areas visible from the public way does not require permission from the owner.

INSPECTION PRIORITIES AND FREQUENCY

Although the model codes mandate the inspection of most structures and premises, they do not establish inspection priorities or frequency. These issues must be determined by the local jurisdiction based on need and available resources.

Inspecting every premises and occupancy on a regular basis is fine if you are employed by a small enough jurisdiction, or one that has the staffing to carry out that many *quality* inspections. But how often is enough? Who needs to be inspected first? Why inspect some occupancies and not others? These important policy issues must be addressed by the senior managers of the fire prevention bureau.

Two models used in determining inspection priorities are the *permit model* and the *inspection model*.[8] Regardless of the system used, it should target potential life safety and property hazards first. It should be fair in that all occupancies within a given classification or type are inspected with the same frequency. Administrative procedures that outline the process should be a written part of the bureau's standard operating procedure.

The Permit Model

■ NOTE
The list of required permits includes occupancies with a high potential for life loss and those involving hazardous processes or the storage and handling of hazardous materials.

Perhaps the simplest model to get started and the easiest to defend as technically valid is the permit model, using the permit requirements and thresholds directly out of the fire prevention code (see Figure 2-1). The list of required permits includes occupancies with a high potential for life loss and those involving hazardous processes or the storage and handling of hazardous materials. Like all provisions of the model codes, they are the result of our nation's fire experience and the collective wisdom of the membership of the model code group. How can you argue against that?

Permits are required to maintain nightclubs because there were fires like the Coconut Grove and the Beverly Hills Supper Club. Buildings housing processes that

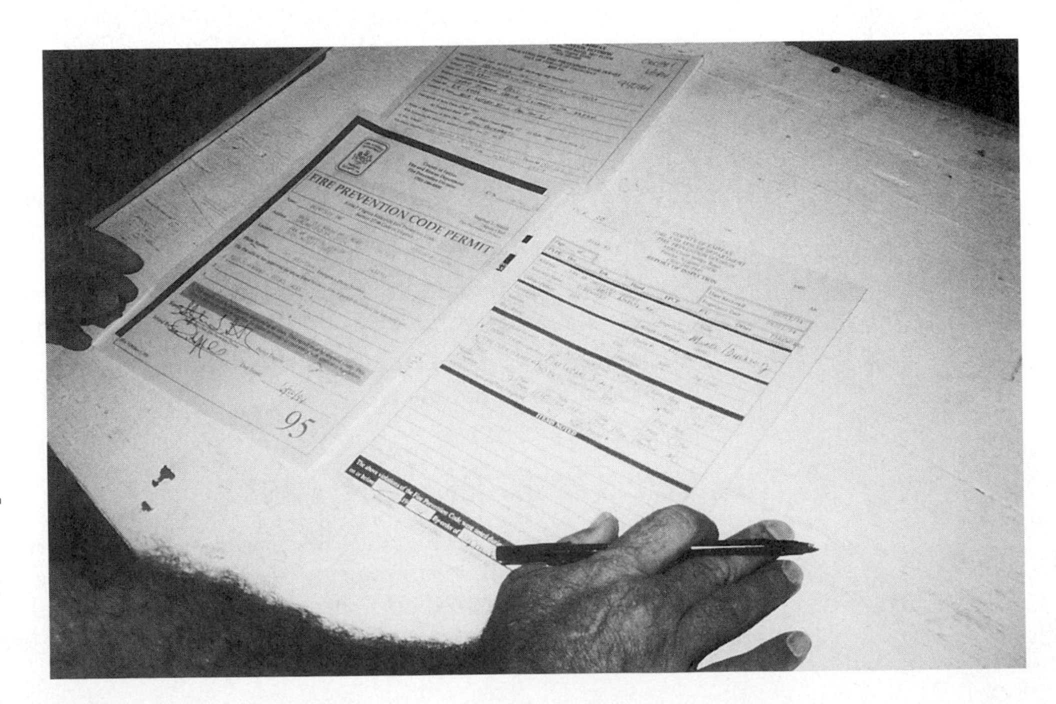

Figure 2-1 *The permit model is easiest to defend on a technical basis. (Courtesy of Lionel Duckwitz.)*

■ NOTE

Permits are required to maintain nightclubs because there were fires like the Coconut Grove and the Beverly Hills Supper Club.

employ flammable liquids and gases and storage occupancies that handle aerosols and other hazardous substances have all been the site of fires that resulted in the loss of life, extensive property damage, and often the catastrophic disruption of the entire community. They all require permits to operate. Those occupancies that have traditionally enjoyed a low incidence of fire, such as business and professional offices, banks, and small mercantile occupancies do not require permits to operate.

The permit process prioritizes inspections for the code official and enables the fire official to identify those occupancies in which fire operations personnel are most likely to encounter hazardous conditions or life hazards. This information should be shared with operations staff. Occupancies requiring permits are those that, in the opinion of the model code groups, have the most potential for fire and life loss.

The burden of application for permit is on the business owner. It is his or her obligation to apply for a permit prior to commencing operations. It is not the responsibility of the fire official to find the business owner. Failure of the business owner to permit inspection at a reasonable hour is grounds for permit revocation.

The permit itself can be a valuable tool for gaining access to structures and premises. Permit fees can be used to help recover the cost of providing inspection services.

The Inspection Model

In the inspection model, the fire prevention bureau determines which types of occupancies will be inspected, identifies them, and conducts the inspections. Written procedures are needed to guard against claims of harassment and selective enforcement by some business owners. If a determination is made to inspect public assemblies during the month of June, then all public assemblies should be inspected throughout the jurisdiction.

Deviation from the schedule is only justified by legitimate circumstances. The failure of several range hood suppression systems in restaurants might warrant special inspections of restaurants outside of, and in addition to, the normal schedule. These can easily be justified based on the actual fire experience of the jurisdiction.

Advantages of the inspection model process include:

- It enables the bureau to pick and choose those occupancy classifications to be inspected.
- It is less labor intensive than the permit process, requiring fewer administrative functions.
- There is more flexibility in the quality of personnel assigned to perform inspections.[9] When using the permit system, inspectors must have the ability to determine compliance with the entire code. Under the inspection model, only one type of occupancy is inspected at a time, theoretically limiting the amount of the code to be applied.

■ NOTE
Using the inspection model, an uneducated or untrained inspector essentially is not noticed until a disaster occurs.

■ NOTE
Better to have *less* inspections, than have *less*-than-adequate inspections.

In *Political and Legal Foundations for Fire Protection*, Howard Markman and coauthors caution code officials about the inspection model, where "using the inspection model, an uneducated or untrained inspector essentially is not noticed until a disaster occurs."[10] Supervisory personnel can be lulled into thinking that all is well because they are not getting requests for help from the inspector and the telephone is not ringing with calls from business owners. The use of untrained or inadequately trained inspections personnel is a recipe for trouble. Better to do *less* inspections, than *less*-than-adequate inspections. The jurisdiction's liability in not performing an inspection at all is questionable unless the statutory duty to perform an inspection is explicit.[11] The potential liability of the jurisdiction, the fire official, and the inspector for a negligent inspection is another matter.

INSPECTION WARRANTS

The Supreme Court has held that warrantless entry by government officials can constitute a serious breach of the Fourth Amendment rights of the occupant of a building. In *Camara v. Municipal Court of San Francisco*, the occupant of an apartment refused to allow housing inspectors to make a routine inspection, as required by city code. He was later arrested for refusing to permit a lawful inspection. In overturning the lower courts, the Supreme Court held that a search of private property is unreasonable unless consented to or authorized by a warrant.[12]

The Court did, however, acknowledge that routine inspection of the physical condition of a property is a less hostile intrusion than a typical policeman's search for the fruits and instrumentalities of a crime.[13] As a result of *Camara* as well as other Fourth Amendment cases, many state legislatures have adopted statutes that establish procedures for obtaining administrative or inspection warrants.

In order to secure an administrative warrant, the inspector must convince an impartial magistrate that either the premises should be inspected based on reasonable legislative or administrative standards or the inspector has reason to believe that a particular violation of the code exists at the premises.

The written administrative procedures that were discussed in connection with the permit and inspections models are critical elements in securing an inspection warrant. There are only two acceptable answers when the magistrate asks why he or she should issue a warrant to inspect a premises. Either you know of a specific violation of the code and require access to conduct a thorough inspection or, based on the written procedures that you can cheerfully provide, the occupancy is due to be inspected.

■ NOTE
It is important to remember that the magistrate represents the people and is supposed to be a neutral party.

It is important to remember that the magistrate represents the people and is supposed to be a neutral party. He or she is not supposed to "be on your side." Don't take it personally if he or she makes you prove your case a little more than you think is necessary. It is the magistrate's job to ensure that your request is based on a legitimate government interest. Do not kid yourself into thinking that *selective code enforcement*, in which certain occupancies or owners are targeted for enforcement efforts for reasons that are not completely legitimate, is not attempted in different communities from time to time.

CONDUCTING THE INSPECTION

Most inspections are conducted with a representative of the owner present and should generally be conducted during the business's nonpeak hours (see Figure 2-2). However, public assembly, such as nightclubs, theaters, and exhibit halls also must be inspected during hours of peak occupancy. If the facility is industrial or occupied by the federal government, you may also be accompanied by safety or union representatives, or both. At times, an owner's representative, attorney, or even a reporter could be on the scene. Remember, *you* are conducting the inspection. *You* are setting the pace and determining the direction to be followed, and *you* should be asking the questions. Do not allow the others, no matter how well intentioned, to distract you. Do not allow the inspection to turn into a sideshow.

■ NOTE
A fair word of advice to almost any inspector is to slow down while conducting inspections.

If you find yourself losing control of the inspection, call the thing off. Come back later, after you have had time to confer with your supervisor, legal department, or just had time to get your act together. Time is on your side, and as you will see throughout this text, you've got all the cards that you need to play the game. A fair word of advice to almost any inspector is to slow down while con-

Figure 2-2 *Routine inspections of restaurants should be conducted early in the day or during slow periods.*

ducting inspections. Talk less, listen more. Ask more questions. Stop and think about what you are seeing. You can not listen with your mouth open.

Ideally, inspections should begin with a review of the records of previous inspections of the facility. That may not be possible, however, especially if responding to a complaint received in the field. A thorough inspection of the outside of the building including fire department access, hydrant accessibility and condition, outside storage conditions, and the viability of the exit discharges should always be attempted *before you even get out of the car*. Drive the long way around the parking lot, or better yet, once around the block.

Take your time and look at the place. Are trucks blocking the fire department connection? Do parked cars block exit doors? How does the outdoor storage of idle pallets and other combustible material affect exiting and building exposures? Complete your inspection of the outside of the facility on foot, after entering the building and meeting with the owner or manager's representative.

NOTE

Your inspection of the inside of a facility should include every room and space within the building.

Your inspection of the inside of the facility should include every room and space within the building. This condition should be made clear at the beginning of the inspection so that the owner or manager's representative brings all appropriate keys along on the inspection. Pick your route through the building at the beginning of the inspection and remember the layout of the structure that you noticed while outside. Many an inspector has been led on a circuitous tour by a building owner and shown only what the owner wanted the inspector to see.

NOTE

The area you might skip because it is inconvenient or difficult to access is probably the same area that the last two inspectors skipped for the same reason.

It is better to ride the elevators up and walk down the steps. Alternating stairwells between floors saves some time. An important point to remember is that the area you might skip because it is inconvenient or difficult to access *is probably the same area that the last two inspectors skipped for the same reason*. Although I am not advocating a full-blown top to bottom, roof to basement inspection of every structure every time, such an inspection certainly should be undertaken at given intervals. This policy issue must be determined on a management level, based on staffing resources, workload, and fire experience within the jurisdiction.

The general condition of the premises also dictates how in-depth the inspection must be. A facility that evidences numerous hazardous conditions or illegal construction must be carefully and thoroughly inspected, including building voids, utility shafts, plenums, and other spaces that might not be carefully examined on every inspection.

Every inspection should include the following minimum components:

- Inspection of the outside of the structure for fire department access, hydrant and fire department connection access and condition, outside storage, and exit discharge viability.
- Inspection of the building fire protection systems to determine they have power, all valves are open and appear to be in general good order, and have current maintenance records. This part includes a trip to the sprinkler room to determine that heat is provided for the sprinkler riser and a trip to the fire pump to verify that the pump is being maintained and run.

- Inspection of the means of egress in the structure, including all interior and exterior exit stairs. Every exit door should be opened and all door hardware should be examined for function and appropriate application. All emergency means of egress lighting should be inspected and tested.
- Inspection of all hazardous processes, hazardous storage, or other special hazards. This part of the inspection is occupancy-specific and can range from chemical hazards in industry to the use of decorative materials in public assemblies.
- Inspection of all mechanical, electrical, and machine rooms.
- Inspection of rated assemblies for penetrations and continuity and detection of construction that appears to be inconsistent with the building construction type.

REPORTING THE INSPECTION

Most fire prevention bureaus have inspection report forms that comply with legal requirements of the jurisdiction. They may advise the building owner of his right to appeal, refer to a section of the municipal code that authorizes the inspection, or even mention the penalties for noncompliance. Philosophies on the subject range the entire spectrum, from no-nonsense heavy-handed to customer friendly. Some are a combination Report of Inspection/Notice of Violation.

Some jurisdictions develop checklist inspection forms. They can be very useful, especially for complex processes and for fire protection system installations. A note of caution is in order however. Checklists should never be considered substitutes for the actual code documents! The model codes and referenced standards are expensive. Over the years, some organizations have attempted to reduce the cost of outfitting inspectors by substituting lists and checklists of code requirements. This is as risky as it is foolhardy. The code documents are the tools of the trade for an inspector. A tradesman who has to borrow tools or who uses less costly and inferior substitutes can never be expected to produce quality work.

The checklist is basically a correlation of code requirements, a user-friendly tool to help the inspector verify specific items regulated by the model codes and referenced standards. Use checklists to guide your inspection and to aid in coordinating the requirements of different standards that may apply to a particular process or system. Be sure to refer to the actual code or standard before issuing a notice of violation, particularly if it involves the court system.

Code officials who caution against using checklists usually list objections such as the checklist becomes a crutch that prevents inspectors from developing proficiency in using the code documents, checklists could be improperly developed and reference incorrect editions of the various codes and standards, and the inspectors have a tendency to check only the items listed on the form and nothing else.

■ NOTE
Every inspection and every visit to the facility should be documented.

■ NOTE
Reports must be legible and clearly state violations noted and corrective action required.

Officials who endorse the proper use of checklists reply that they are simply additional tools for the well-trained and well-equipped inspector. Both sides cite valid concerns. These issues are as old as the trade itself. There has never been a substitute for good training and proper equipment, in any trade or craft.

Every inspection and every visit to the facility should be documented, including visits where access is denied or the inspection rescheduled at the request of the owner. These documents may be critical elements in obtaining an inspection warrant in the future. Reports must be legible and clearly state violations noted and corrective action required. By merely looking at your report, will the judge understand what you were ordering the building owner to do? If your correction order is not clear and concise enough to pass this test, you could be 100 percent right and walk out of court the loser.

VIOLATIONS

Notices of Violation

Written notice to the owner or occupant of a premises is required by the model codes:

■ NOTE
The use of the term *shall* indicates mandatory actions or requirements under the law.

■ NOTE
If the word *may* is used, discretion is allowed.

■ NOTE
The issuance of a notice of violation in the proper legal form is a mandatory requirement of the code.

> *Whenever the code official observes an apparent or actual violation of the provision of this code or other codes and ordinances under the code officials jurisdiction, the code official shall prepare a written notice of violation describing the condition deemed unsafe and specifying time limits for the required repairs or improvements to be made. . .* (*F-1121 1999 BNFPC*, p. 7)

> *When the Chief finds any building or premises . . . that this code is being violated, the chief is authorized to issue such orders as necessary for the enforcement of the fire prevention laws and ordinances governing the same and for the safeguarding of life and property from fire. 103.4.1.1 . . . orders or notices which are given verbally shall be confirmed by service in writing as herein provided.* (103.4.2 *2000 UFC*)

> *If during the inspection of a building, the building or any of the building systems in whole or in part constitutes a danger to human life, or a hazard to safety or health the fire official shall issue such notice or orders to remove or remedy the conditions . . .* (102.4.1 *1994 SFPC*)

> *Whenever the authority having jurisdiction determines violations of this Code, a written notice shall be issued to confirm such findings.* (1-19.1 NFPA 1, 2000)

> *When the code official finds a building, premises, . . . that is in violation of this code, the code official is authorized to prepare a written notice of violation . . .* (102.9 2000 *IFC*)

shall
indicates a positive and definitive requirement of the code that must be performed; action is mandatory

may
indicates a discretionary provision of the code; enforcement is left to the judgment of the fire official

negligence
culpable carelessness; the failure to act as a reasonable and prudent person under similar circumstances

The use of the term *shall* indicates that these are mandatory actions or requirements under the law. If a law or code states that something **shall** be done it is a "positive and definitive requirement. If the word **may** is used, discretion is allowed."[14] The issuance of a notice of violation in the proper legal form is a mandatory requirement of the code. The *BNFPC* and *NFPA 1* give the fire official no option when he observes a violation of the code. He must issue a written notice. The *UFC* "authorizes" the fire official to issue orders or notices either verbally or in writing, but requires verbal notices to be confirmed by written notice.

In 1970, five persons perished in a fire at the Gold Rush Hotel in Anchorage, Alaska, eight months after the state fire marshal's office had inspected the hotel and detected numerous code violations. The lawsuit alleged that the state fire marshal's office had promised, but failed to provide the owner a letter with required corrective actions needed. The Alaska Supreme Court upheld a lower court ruling of **negligence**, in that the fire official's duty to provide the written corrective order in this case was not discretionary, but mandatory.[15]

The model codes also include specific methods of delivery, which include personal delivery, delivery by mail to the last known address, and the actual posting of the structure. Extreme care must be taken to ensure that all appropriate requirements are complied with. Personal delivery is always the most desirable method, followed by certified mail. Posting the structure is least desirable and invariably leads to claims in court by the owner that he "certainly would have complied if he had only known."

■ NOTE
Personal delivery of notices of violation is always the most desirable method, followed by certified mail.

Elements of a Notice of Violation Notices of violation that follow a prescribed format, such as a fill-in-the-blank printed form are preferable to a letter. The form can be preapproved and used by the entire bureau. As a minimum, notices of violation should:

- Be in writing and include the date, legal address of the premises inspected, and name of the inspector.
- State the specific violations noted by the inspector and identify the applicable code section for each violation.
- State the required corrective action.
- List the date on which a follow-up inspection will be performed to ensure compliance.

■ NOTE
You cannot charge somebody else with a code section that requires you to act.

GAINING COMPLIANCE

Enforcement actions for noncompliance depend on your state law, the local ordinances of your jurisdiction, and the policies of your local government and department.

Criminal Charges

summons
a written order, issued by a judicial officer, law enforcement officer, or other authorized official, directing an alleged offender to appear in court at a specific time to answer a criminal charge; a summons issued by a law or code enforcement officer is classified as a *citation*

citation
a written order, issued by a law enforcement officer or other authorized official, directing an alleged offender to appear in court at a specific time to answer a criminal charge

injunction
a legal order issued by a court that commands a person or entity to perform a specific act or prohibits a specific action by that person or entity

In some jurisdictions, violations are criminal misdemeanors and inspectors are empowered to issue **summonses** or **citations** for violations of the fire prevention code. This method is effective in that it quickly *gets the owner's attention.*

There is a downside however. You are bringing criminal charges here, this is not traffic court. Many prosecuting attorneys have little knowledge of fire or building codes and even less time or interest in learning about them. If you want to win your case, you better do your homework and provide the prosecutor with a succinct, one-page script or summary. Include the sections of your state, county, or municipal code that adopt the model code, the sections that empower you to enforce it, and the specific section of the fire prevention code that has been violated.

It should be noted here that the *Unsafe Conditions* sections (*BNFPC*-110, *UFC* Sec. 103.4.1.2, *SFPC* 102.2, *NFPA 1*, 1-4.5) contained within some of the codes require action on the part of the *fire official* in ordering the remedy of certain unsafe conditions in accordance with the provisions of the rest of the code. They are not blanket catch-all sections to be used to charge code violators. You can not charge *somebody else* with a code section that requires *you* to act.

Civil Charges

An **injunction**, or order from a court with jurisdiction prohibiting a specific action (in this case a code violation), can be effective in gaining compliance. If the violation continues, the violator is in contempt of court and can be jailed until compliance is attained.

The downside is that the preparation of the motion for injunctive relief must be prepared by your jurisdiction's legal department, and a lawyer must present the case in court. This is an extremely drawn out, labor-intensive process. Some cases become so political or so complicated that injunctive relief is the remedy of choice.

In some jurisdictions, civil penalties can be assessed for code violations, often without taking the violator to court. Civil fines often act as a deterrent, but to some, they are merely the "cost of doing business."

APPEALS

■ **NOTE**
Failure to comply with any of the provisions of proceeding with an appeal are possible grounds for the dismissal of the appeal.

The appeals process is a two-edged sword. Use it wisely and it can aid in your code enforcement efforts and keep the code official out of a lot of hot water. Abuse it or fail to respect the potential damage it can cause and it will become a nightmare come true. The elements of the process, including who may file an appeal, the application process, and the time limits for each step are clearly prescribed by the codes or within adopting ordinances. Failure to comply with any of the provisions are possible grounds for dismissal of the appeal if the appellant has violated the process, or a finding against the code official if he has violated the

■ NOTE
Boards of appeals are responsible for hearing appeals with the technical aspects of the code.

process. At the very least the local official who can not follow his own process gives the jurisdiction a black eye.

Appeals Boards

Each of the model fire prevention codes has different provisions for the makeup of the boards. The *BNFPC* requires that the board consist of five registered design professionals within the different building and fire protection disciplines. The IFC establishes the board in the administrative provisions of Chapter One. Additional information on appeals boards is included in Appendix A of the IFC. Appendices are generally informational and are not adopted as part of the code. The *UFC*, *SFPC*, and NFPA 1 give the local jurisdiction flexibility in determining the makeup of the board.

Some jurisdictions include members of the real estate industry, attorneys, and citizens at large. Often such appointees lack the technical expertise in the code to render competent decisions and fall back on their political or social beliefs. This is unfortunate and should be avoided.

In some jurisdictions the same board may hear appeals from both the building and fire prevention code official's decisions. It is very important that both the makeup of the board and the scope of their responsibility be clearly stated by the local governing body when the board is constituted. Boards of appeals are responsible for hearing appeals dealing with the *technical* aspects of the code, not with the political or philosophical movements that may have led to code development.

Appeals boards cannot and should not debate the validity of a portion of the code. That is for the local governing body or the courts to decide. The zealot, with a political ax to grind, has no place on an appeals board. The citizen whose personal conviction that less government regulation is better or that government needs to rectify some past social wrong and casts his vote based on political belief, is a danger to the safety of the public and the code process. Such behavior undermines the code process.

Appeals Hearings

Hearings are public meetings that must comply with state and local laws that regulate government business. Hearings are generally advertised in newspapers or government agendas with public distribution and held in public buildings. The proceedings are recorded or transcribed and become public record.

Although hearings are structured and run by the chairman, they are generally informal enough to ensure that all parties have an adequate opportunity to present their case. The chairman is responsible for ensuring that the hearing remains on track and does not become a session of bullying the public officials involved. Once arguments have been made and rebutted by both sides, members of the board pose questions to the appellant and code official before deliberating and rendering a decision. Appeals of that decision may go to a state board or to the courts.

Hearing Preparation

"It is this way because we're the fire department and we say so," won't get you very far before the appeals board. The fire department "mom and apple pie" is a great angle, and you should make the most of it, but you must win the appeal on the merits of the issue.

Cooperation between building and fire officials is critical when it comes to the appeals process. When the building official testifies that an illegal locking arrangement is unsafe because it is a code violation, the board gets one message. If the next witness is a fire officer in dress uniform who testifies that the locking arrangement is unsafe because it precludes the use of required exit doors in the event of an emergency, potentially hampers fire department operational efforts, and potentially endangers the building occupants, your one-two punch may save the day.

NOTE
Cooperation between building and fire officials is critical when it comes to the appeals process.

When appearing before the board:

- Provide board members all the documents involved in the appeal as well as a brief one-page synopsis of the issues.
- Have adequate, well-prepared handout material for all board members as well as the appellant and members of the public that may be present.
- Your dress, speech, and mannerisms are crucial elements of your presentation at the hearing. It may be advantageous to have several witnesses for the jurisdiction, such as the fire marshal in civilian dress, a fire officer who will speak to operational issues in dress uniform, or even a fire investigator. The board probably will not understand your department's organizational structure, so use it to your advantage.
- Be cordial and polite to the members of the board, the appellant and his attorney or representatives, and any witnesses present. A handshake and hello *before* the hearing and *at the conclusion* will help set the tone for nonhostile hearings that stick to the technical issues.
- Do not take the process or the results personally. Sometimes logic seemingly goes out the window at such hearings. Critique your presentation and make the next one better.

RECORDS

A fire prevention bureau lives and dies by its records. Reports of inspection, notices of violation, code modification letters, and code interpretations are but a few of the documents that must be maintained. Most states have retention schedules for public documents that mandate the storage of certain records for fixed periods of time. The *Uniform Fire Code* requires that inspection and investigation reports be maintained for three years. Failure to comply is a violation of state statute or local ordinance, as well as an embarrassment. You may be violating your own code.

FREEDOM OF INFORMATION AND PUBLIC ACCESS LAWS

Freedom of Information Act
a law passed by Congress and signed by President Lyndon Johnson in 1967 that, as well as the state laws patterned after it, guarantees public access to all documents and information under the control of government, with certain specific exemptions

Most states have adopted laws that guarantee public access to the workings of government. Many of these were patterned after the **Freedom of Information Act** signed by President Lyndon Johnson in 1967.[16] The "public" had a lot less to do with the passage of these laws than the press did, but the laws have opened the doors on many government functions. Every person who holds a government position, whether compensated or not should have a clear understanding of these regulations and the impact on their duties.

■ NOTE
Generally, all documents in the possession of government are public records.

Generally, all documents in the possession of government are public records, and as such must be made available for public inspection and copying. There are specific exceptions such as personnel records, medical records, ongoing criminal investigations, and others, depending on your state statute. The federal statute permits exemptions in only nine specific areas including national defense, law enforcement files, certain personnel files, and other information whose disclosure is prohibited by statute.[17] A copy of the law and guidelines for compliance are available from some state attorney general's offices. Check with your jurisdiction's legal officer.

■ NOTE
Requests for information are often subject to very strict time constraints.

Requests for information are often subject to very strict time constraints in which government officials must respond within a given number of days or risk court sanctions. A response that the records are not available in the form requested may be adequate, if it is made within the allotted time frame. Freedom of information laws:

- Make documents, recordings, videos, and other media in the possession of government open for public inspection. Exceptions include ongoing criminal investigations, medical and personnel records, and attorney–client communications.
- Require government to respond within a given time frame to requests for information.
- Require that all meetings of boards, commissions, and elected bodies be open to the public and advertised in advance.
- Do not require government agencies to create documents such as studies, comparisons, or summaries in response to requests. Requests are limited to records that exist in their current form.

■ NOTE
Freedom of information laws contain specific penalties for government officials who fail to comply.

These laws contain specific penalties for government officials who fail to comply, and they can be quite stiff. Review the law in effect in your state. If your agency does not have a written standard operating procedure for requests for information, it is time to get one. The first time you face the judge to explain why a reporter was denied access to an inspection report, you'll wish you had.

Summary

The inspection of buildings and premises is the real backbone of a fire prevention program. The fire prevention code drives the inspections process by establishing the scope of the process, right of entry to inspect, and actions to be taken when noncompliance is discovered.

All structures fall within the scope of the code, although one- and two-family dwellings and the dwelling units of multifamily dwellings are exempt from routine inspection. Administrative or inspection warrants are required to gain access to residential and commercial property when entry is denied.

Two methods of determining inspections priorities are the permit model, in which permits are required for certain hazardous processes and occupancies, and the inspection model, in which certain occupancy types are inspected throughout the jurisdiction at a given time.

Inspections should begin with the outside of the structure or premises. Every inspection should include the building fire protection systems, viability of all portions of the exits, fire protection elements of the structure, building utilities, hazardous processes, and issues affecting the life safety of the occupants. Every inspection should be documented on the proper form.

The preparation and issuance of notices of violation are a mandatory requirement of the model fire prevention codes. Specific provisions in the case of noncompliance vary according to locality. Criminal charges, civil actions, or both are available to some jurisdictions.

Appeals from decisions of the fire official are heard by boards established by the model codes in similar processes. The makeup of the boards as well as the scope of their responsibility and authority are crucial to their effectiveness. Fire officials must carefully prepare for hearings and should coordinate their efforts with the other code officials of the jurisdiction.

Various freedom of information laws have been passed by the states which guarantee public access to records in the possession of public agencies and require government business to be conducted in the public forum. The laws contain penalties that the court may impose if government officials are found guilty of violating the statutes. Every bureau should have a copy of the statutes in effect within the jurisdiction, as well as written standard operating procedures for handling public requests for information.

Review Questions

1. A ______________ clause is a code provision that mandates the inspection of certain occupancies.
2. The right of entry to inspect a premises is limited by the ______________ Amendment of the Constitution.

3. When denied entry to perform an inspection, what remedy is available to the fire prevention bureau? ________________

4. Name two administrative methods of determining inspection priorities.
   1. ____________________
   2. ____________________

5. The use of the term ________ indicates that a provision is mandatory, and discretion on the part of the fire official is not allowed.

6. As a minimum, notices of violation should consist of what four elements?
   1. ____________________
   2. ____________________
   3. ____________________
   4. ____________________

7. When a court with jurisdiction issues an order prohibiting a person from violating the fire prevention code, this order is called a(n) ______________.

8. Appeals boards should render decisions based on
   A. political considerations.
   B. economic considerations.
   C. technical merit.
   D. social justice.

9. Name three types of records that are generally exempt from disclosure under state freedom of information laws.
   1. ____________________
   2. ____________________
   3. ____________________

10. Are government agencies obligated to create reports or summaries of information when requested under state freedom of information laws? __________

Discussion Questions

1. Your office has received a complaint from a citizen who claims that her neighbor is storing a 55-gallon drum of racing fuel in the basement of his town house. Upon questioning she accurately describes the red flammable liquid label on the drum. Does the fire prevention code apply in this situation? If denied entry by the homeowner, what options are available to gain entry? What effect could the complainant's desire to remain anonymous have on your ability to gain access to the town house?

2. A local fire protection contractor has requested copies of inspection reports for a list of specific occupancies that cancelled service contracts with his company in the preceding year. Is he entitled to these records under the freedom of information law?

Section

2

THE BUILDING CODE

The model building codes are lengthy documents—they have to be in order to regulate everything from concrete aggregates and elevator hoistway doors to accessibility for the physically disabled. Do not let that bother you. The fire inspector needs a good working knowledge of about one-third of the building code and the ability to find his way around the rest.

However, unless you understand how the egress facilities and fire protection features of buildings are designed and constructed, you cannot possibly ensure that those features are intact and functional during your inspection. Knowledge of building design and construction is crucial for a competent fire inspector. It will also help you, the fire inspector, develop a professional working relationship with the building department. Professional expertise is respected in every profession.

Use and Occupancy

Learning Objectives

Upon completion of this chapter, you should be able to:

- Describe the classification system used by the model building codes to group buildings by use and occupant characteristics.
- Describe how multiple uses or *mixed uses* within a single building are addressed by the model building codes.
- Explain the need for *special occupancy requirements* in some buildings.
- Explain the fire official's responsibility in illegal changes in use.

■ Note
The use group classification is the first aspect that must be established in the design of a building and should also be your first consideration during an inspection of a building.

use group
building code classification system whereby buildings and structures are grouped together by use and by the characteristics of their occupants

Open your building code to the table of contents. In all codes but NFPA 5000, the first two chapters are largely administrative procedures, definitions, and housekeeping. In NFPA 5000, the first five chapters include definitions, administrative procedures, referenced standards, and design criteria. Technical aspects of the code, or the nuts and bolts of designing a building, begin with use group classifications in Chapter 3 of the *IBC, BNBC, SBC,* and *UBC,* and Chapter 6 of NFPA 5000. The use group classification is the first aspect that must be established in the design of a building. It should also be your first consideration during an inspection of a building.

The **use group** classification of a building is probably the most significant single design factor that affects the safety of the occupants as well as the fire suppression forces that are called on in the event of fire. The building's height and size, type of construction, type and capacity of exit facilities, and fixed fire suppression systems are all dependent on this classification. The system of use group classification is the foundation for the building and fire prevention codes. As we will see, a mistake in laying the foundation can prove to be a disaster later on. Failure to maintain the foundation can also be disastrous.

One of the primary responsibilities of the building official is to verify, during the plan review process, that the design professional has accurately defined the use and occupancy of the building and that the design of the building is in accordance with the code requirements for that use. The fire official's responsibility for the life of this same building is to determine that it is still operating under that specific use, and that the building protection features including means of egress, fire-resistance rated assemblies, and fire protection systems are being maintained. The fire official has the added responsibility of ensuring that functions and processes occurring within the facility are in accordance with the fire prevention code.

USE GROUPS

The model codes as well as NFPA 101 separate buildings into about ten general uses (Table 3-1). The uses are further separated by specific characteristics into *use groups.* A church, a nightclub, and a family restaurant are all assemblies, but the specific characteristics of their occupants and functions differ drastically, requiring different built-in levels of protection. The occupants of the church are probably very familiar with the building that they occupy. They have been there before and know the location of alternate exits. The occupants of the nightclub may not be so familiar. Dim lighting, loud music, and impairment by alcohol are all common features that may affect the ability of the occupants to identify a fire emergency and take appropriate measures to escape.

Assembly (A) occupancies are subdivided by function as well as the number of occupants. Assemblies capable of holding fewer than fifty persons generally are

Table 3-1 *Use groups.*

IBC, BNBC, SBC, UBC	NFPA 101, NFPA 5000
A—Assembly	Assembly
B—Business	Business
E—Educational	Educational
F—Factory or industrial	Industrial
H—High hazard	Health care
I—Institutional	Mercantile
M—Mercantile	Residential
R—Residential	Storage
S—Storage	Detention and correctional
U—Utility, miscellaneous, or special	Ambulatory health care
	Residential board and care

considered as less restrictive business uses. The *Uniform* and *Standard Building Codes* further subdivide assemblies that hold many people. Assembly areas include churches, restaurants with occupant loads that exceed fifty persons (100 under the *SBC*), auditoriums, armories, bowling alleys, courtrooms, dance halls, museums, theaters, and college classrooms holding more than fifty persons (100 under the *SBC*).

Business (B) areas include college classrooms with occupant loads under up to fifty (100 under the *SBC*), doctors' and other professional offices, fire stations, banks, barber shops, and post offices. Dry cleaners who use noncombustible solvent (types IV and V) are Business uses.

Educational (E) areas include areas *not* used for business or vocational training (shop areas), for students up to and including the twelfth grade. Colleges and universities are Business or Assembly areas depending on the number of occupants. *Day care facilities may be classed as Educational or Institutional depending on the model code.*

Factory and Industrial (F) areas include industrial and manufacturing facilities and are subdivided into moderate- and low-hazard facilities. High-hazard factory and industrial areas are "bumped" up from the F Use Group to the H Use Group. Dry cleaners employing combustible solvents (type II and III) are Moderate Hazard Factory and Industrial uses.

High Hazard or Hazardous (H) areas are those in which more than the exempt amount of a hazardous material or substance is used or stored. *Exempt amounts* of hazardous materials are not exempt from the provisions of the code. They are a threshold amount by material, above which the occupancy must comply with the stringent requirements of Group H use.

Institutional (I) areas may include halfway houses and group homes, hospitals and nursing homes, and penal institutions. The model codes differ in their breakdown. Care must be taken when considering homes for adults and day care centers as to whether the occupants are **ambulatory** or "capable of self preservation." The model codes all contain significantly more stringent requirements for institutional occupancies where a "defend in place" strategy is necessary due to the inability of the occupants to flee the structure without assistance.

ambulatory
able to walk about without assistance; capable of sensing an emergency situation and appropriately responding by exiting the building

Mercantile (M) uses are retail shops and stores and areas displaying and selling stocks of retail goods.

■ Note
The model codes all contain significantly more stringent requirements for institutional occupancies where a "defend in place" strategy is necessary.

Residential (R) areas are hotels and motels, dormitories, boarding houses, apartments, town houses, and one- and two-family dwellings.

Storage (S) areas are used for the storage of goods and include warehouses, storehouses, and freight depots. Storage uses are separated into low- and moderate-hazard storage uses. Auto repair facilities that perform major repairs, including engine overhauls and body work or painting are considered Moderate Hazard Storage occupancies by the *BOCA* and *Standard Building Codes*, and Hazardous by the *Uniform Building Code*. Occupancies that store over the exempt amounts of hazardous materials or substances are considered Group H.

Utility (U), Special, or Miscellaneous, depending on the model code are those not classified under any other specific use, such as tall fences, cooling towers, retaining walls, and tanks.

Although NFPA 5000 uses slightly different classifications, the same basic principles apply. Occupancy classifications from the building codes are based on the relative hazards of three conditions that exist in buildings: characteristics of the occupants for self-preservation, processes that take place in the building, and types and amounts of regulated materials within the building. NFPA 5000's use classifications for Ambulatory Health Care, Detention and Correctional, and Residential Board and Care are included as subgroups within the use groups used by the other codes. NFPA 5000 regulates occupancies containing hazardous materials as "hazardous contents" within one of the other use groups. The other model codes classify these buildings as one of the five high-hazard uses.

MIXED USES

Buildings often contain multiple occupancies (see Figure 3-1) with different uses. The three-story office building might have a restaurant (assembly) and computer store (mercantile) on the first floor and professional offices throughout the rest of the building. The model codes provide for these situations either by requiring that the whole building be constructed to all the requirements of the most restrictive use group, or by separating the areas with fire-rated assemblies, or by separating the building with fire walls. Separating with fire walls actually creates separate buildings. By far the cheapest, most attractive method of separating mixed uses is by using fire separation assemblies, but this method is sometimes impossible due to building height and area requirements, which we discuss in Chapter 4.

accessory use area
a portion of a building with a different use group classification than the main area, but which does not require a fire separation between the main area, due to its small size

The *2000 International Building Code* incorporates a concept taken from the *BOCA National Building Code*, Accessory Use Area. An **accessory use area** is a portion of a building with a different use group classification than the main area, but which does not require a fire separation between the main area, due to its small size. In order to be considered an accessory use area, it must not be a High Hazard use, and shall not exceed 10 percent of the floor area of any floor, and meet requirements for height and area based on construction type. There is no definition within the *IBC*, which may lead to some confusion for previous users of the *Standard Building Code* and the *Uniform Building Code*, but the concept worked well in the *BOCA National Building Code* for years without a definition.

Figure 3-1 *The typical "B" use building is actually a mixed-use building. (Courtesy of Duane Perry.)*

Once the use group has been determined, the code becomes a much friendlier document. Requirements that range from building height to the need for fixed fire protection systems are all based primarily on the use group of the building. The code writers found that basing these requirements on the intended use or the nature of the occupants of the building worked very well, with a few exceptions. Those few exceptions make up an entire chapter in your building code—Special Use and Occupancy.

SPECIAL USE AND OCCUPANCY REQUIREMENTS

■ Note

Building codes provide an enhanced level of protection for certain occupancies to compensate for special hazards, over and above those posed by the use of the building alone.

For most buildings and structures, assigning a use group and then specifying building requirements for all buildings within that use group works relatively well. Most mercantile occupancies share common hazards. Most business occupancies have similar occupants and processes. But what if that business happens to be on the twenty-sixth floor of a high-rise building? Or what if the men's clothing store is in the middle of a giant shopping mall? The relative hazards suddenly change, and we begin comparing apples to oranges.

Building codes provide an enhanced level of protection for certain occupancies to compensate for special hazards, over and above those posed by the use of the building alone. The inherent hazards posed by being located twenty-six stories above the ground (Figure 3-2), or in a large open area with high fire loading

Figure 3-2 *High-rise buildings under construction.*

such as a shopping mall, are addressed as *special use* requirements. The list of special uses seems to grow with each new edition of the model codes.

Covered Mall Buildings

An understanding of the concept of a shopping mall is important. In your building code are definitions for *covered mall building* and *anchor store* or *anchor building*. As with all of our subjects, read and understand the definition! A covered mall is considered a building that encloses a number of tenants and occupancies such as retail stores, amusement facilities, restaurants, offices, and even transportation terminals where two or more tenants have a main entrance into the mall.[1]

Anchor stores, which are usually the large department stores, are *not* considered to be a part of the mall. The required means of egress from anchor stores are totally independent from the mall. Exits leading directly to the building exterior must be provided, as if that huge opening into the mall were not there. The occupancies that are part of the mall exit through the mall, but travel distance through the exit access is limited to 200 feet. Those back corridors that mall management refer to as "service and delivery" are actually the exits for the mall shops. That 200 feet of travel distance is to the entrance to the back corridor, *not to the outside*. The merchants who stack empty boxes (see Figure 3-3) there, or block them with carts, or prop the doors open should be getting a visit from you. Nothing is permitted to be stored in the exit, period.

Covered mall buildings have special requirements for egress, fire separation of tenants, fire suppression, and smoke control, over and above those of the use groups of the tenants.

Figure 3-3 *Service corridors in malls are actually exits. This exit has been illegally blocked by storage. (Courtesy of Ron Berry.)*

High-Rise Buildings

High-rise buildings (see Figure 3-4 and 3-5) are those with occupied floors more than 75 feet above the lowest level of fire department access. The code attempts to address the practical difficulties faced by the occupants in fleeing the structure, as well as those faced by the fire department in manually extinguishing a fire. These include the use of noncombustible construction materials, the installation of sprinklers, standpipes, and a fire command station with control of various building systems. Stairwell pressurization, selective alarming of floors, and even stairwell locking requirements are all special requirements above and beyond those of the building's use group.

Special Amusement Buildings

Special amusement buildings range from fun houses at carnivals to laser tag facilities in strip shopping centers. The common element is that the means of egress

Figure 3-4 *Early high-rise construction lacks many of the safeguards found in modern high-rise buildings.*

Figure 3-5 *Modern high-rise construction includes sprinkler systems, pressurized stair shafts, and fire command stations.*

is intentionally confounded through the use of a maze, movable partitions, darkness, sound effects, or a combination of all of them. Without special precautions, emergency egress would be severely impacted. The codes require the automatic restoration of adequate lighting and the stopping of special audio effects upon automatic smoke detection or activation of the sprinkler system. Constant attendance at a station with manual sound and lighting controls is also required. All decorative material must also be noncombustible.

Other Special Uses

Special requirements for building atriums, high-hazard occupancies, hazardous production material facilities (HPM), and institutional occupancies are also included in this chapter of your building code. For these limited situations, the use group classification system that works so well everywhere else in the code, did not quite fit the bill. The code writers found they needed to add a little extra.

CHANGES IN USE

Detecting and prohibiting illegal changes in use is a primary responsibility of the fire inspector. Some changes in use do not appear that drastic on the surface, but some pose immediate hazards to the occupants. The 10,000-square-foot

Note
Illegal changes in use are a violation of the fire prevention code.

unsprinklered retail store that turns into a bar by magic overnight might not seem like that big a change. At least not until you go out to the car and check the building code. The threshold for requiring an automatic sprinkler system in a retail store is 12,000 square feet, so the building was not sprinklered when it was originally built. Under the *IBC*, the threshold for requiring sprinklers in a bar or tavern is 5,000 square feet. The lack of sprinklers is not the only potential problem. The building will require panic hardware on the exit doors, a fire alarm system, additional exits, and more.

Illegal changes in use are a violation of the fire code. You have the responsibility and authority to prevent the illegal change. What you as the fire official or fire official's assistant *do not* have the authority to do under the model codes is to order the required upgrades and then approve the new use. Our retail store/bar example above, clearly demonstrates the need for the fire official to work with the building official. Although the IFC authorizes both the fire official and building official to require the installation of sprinklers, that may not be the only deficiency requiring an upgrade to meet the requirements for the new use.

Changes in use require a complete analysis of a building prior to approval. It is as if the owner is building a new building on the site. The fire inspector who orders the installation of a sprinkler system and then finds out later that the building is too big, or built of the wrong type of materials, or built too close to the lot line exposes himself and his jurisdiction to liability and public scorn, not to mention damaging the working relationship between the building and fire departments.

Summary

■ Note
No single other design element has a greater effect on the safety of the occupants than use group classification.

No single other design element has a greater effect on the safety of the occupants of a building than use group classification. The use of a building determines the number of occupants as well as their physical and mental characteristics. Their capacity for self-preservation as well as the physical hazards they will be exposed to are all addressed through the use group classification system. Special hazards and circumstances are addressed through *special use and occupancy requirements.*

The code builds on this classification, by regulating the height and area, type of construction, separation, and fire protection features of buildings based on use group.

Review Questions

1. In designing a building, the first and most critical step is to determine the ___________.
2. Buildings in which multiple uses occur are called ________________.
3. Describe three methods used by the model codes to address buildings in which multiple uses occur.
   1. ____________________
   2. ____________________
   3. ____________________
4. Name three types of buildings that have special use and occupancy requirements.
   1. ____________________
   2. ____________________
   3. ____________________
5. A(n) ________________ is a large store on the exterior perimeter of a covered mall that has exits independent of the mall.
6. List three occupancy characteristics of a tavern or a nightclub that cause it to be classified as its own use group.
   1. ____________________
   2. ____________________
   3. ____________________
7. List three occupancy characteristics of a house of worship that contrast with those listed in question 6.
   1. ____________________
   2. ____________________
   3. ____________________
8. What is the principal occupant characteristic that results in different use group classifications for nursing homes and board and care facilities? ____________________
9. Are illegal changes in use a violation of the model fire prevention codes? _____
10. Does a fire inspector have the authority to order the installation of a sprinkler system in a building that has been illegally changed from a warehouse to a nightclub? _____

Discussion Question

1. You are called by the administrator of an adult care facility classed as I-1 (*BNBC* or *IBC*), R2 (*SBC*), or Group I Division 2 (*UBC*). She requests your assistance in updating the fire safety plan for her facility, because most of the occupants are in wheelchairs and the old plan called for total evacuation. She has drafted a new plan using the defend in place strategy in use at the last facility she worked at and would like your approval.

   a. Does the current occupancy reflect the use group assigned to the building?

   b. How does the use group affect fire safety and evacuation planning?

   c. Is the defend in place strategy proposed by the director compatible with the built-in fire safety features of the building?

Building Limitations and Types of Construction

Learning Objectives

Upon completion of this chapter, you should be able to:

- Define *fire resistance*.
- Describe the five basic construction types.
- Explain why building height and area are regulated by the building codes.
- Describe the methods used by the codes to permit height and area increases.

■ NOTE
The largest part of the model building codes addresses fire protection issues.

Two of the most effective methods used over the years to limit potential fire spread and prevent conflagration have been limiting the size of buildings and regulating the materials used in construction. The same principle behind the city of London's 1189 ordinance requiring that all new buildings be constructed of stone walls with tile roofs[1] is used today in our modern building codes.

One of the primary purposes of a building code is to prescribe standards that will keep buildings from falling down. Besides gravity, there are many forces that act against a building. Snow loads, wind loads, and potential earthquake loads are all provided for in the design and construction of a building by the building codes. The potential force that requires the most extensive code provisions is fire. The largest part of the model building codes addresses fire protection issues, fire safety, emergency egress, and structural stability when exposed to fire.

fire resistance
the resistance of a building to collapse or to total involvement in fire

The key to understanding building code provisions for structural protection from fire is the concept of **fire resistance**. In broad terms it is the resistance of a building to collapse or to total involvement in fire, and has also been called *fire endurance*. It is measured by the length of time typical structural members and assemblies resist specified temperatures.[2]

fire resistance
property of materials and their assemblies that prevents or retards the passage of excessive heat, hot gases, or flames under conditions of use

The term **fire resistance** is defined in the building codes as: *That property of materials or their assemblies which prevents or retards the passage of excessive heat, hot gases or flames under conditions of use.*[3] Fire-resistive construction elements, including methods used to test and assign relative values to their performance in hours, are discussed in Chapter 5.

TYPES OF CONSTRUCTION

We tend to make the subject of construction type much more difficult than it really is. There are three key points to remember when dealing with building construction types:

protected
shielded from the effects of fire by encasement in concrete, gypsum, or sprayed-on fire-resistive coatings

1. All construction is either combustible (it will burn) or noncombustible (it won't).
2. "**Protected**" when applied to construction materials means protected from the effects of fire by encasement. Concrete, gypsum, and spray-on coatings are all used to protect construction elements. *When the code means "protected with a sprinkler system," it will say just that.*
3. Having the ability to eyeball a building and determine the construction type is not a requirement of the job, and in fact will often be impossible for the inspector in the field—accept this. It is easier to determine *what it is not than what it is*, and usually that gets the job done.

■ NOTE
"Protected" when applied to construction materials means protected from the effects of fire by encasement.

An inspector needs the ability to recognize the questionable presence of combustible construction in a noncombustible building, or the apparent damage or compromise of rated components or assemblies.

Five Construction Types

Five basic construction types are recognized by the model building codes. The *Standard Building Code* subdivides noncombustible construction and uses six types. The terms vary a little between the different codes, but the concept is the same. Based on the classification from the *International Building Code* (Table 4-1).

Type I: Fire Resistive In type I construction (see Figure 4-1), the structural elements are noncombustible and are protected. Type I is divided into two or three subtypes depending on the model code. The difference between them is the level of protection for the structural elements (expressed in hours). Only noncombustible materials are permitted. Structural steel will not be exposed. A high-rise building with a steel structure encased in concrete is an example of a type I building.

Type II: Noncombustible In type II construction (see Figure 4-2), the structural elements are noncombustible or limited combustible. Type II is subdivided into subtypes, depending on the level of protection (in hours) for the structural elements. The buildings are noncombustible, but are afforded limited or no fire resistance for the structural elements. A strip shopping center, with block walls, steel bar joists, unprotected steel columns, and a steel roof deck is a type II (unprotected) building.

Type III: Limited Combustible (Ordinary) In type III construction (see Figure 4-3), the exterior of the building is noncombustible (masonry) and may be rated depending on the horizontal distance to exposures. The interior structural elements may be combustible or a combination of combustible/noncombustible. Type III is divided into two subtypes, protected and unprotected. The brick, wood joisted buildings that line our city streets are of type III or "ordinary construction." *Buildings with a masonry veneer over combustible framing are not type III.*

Type IV: Heavy Timber In type IV construction, the exterior walls are noncombustible (masonry), and the interior structural elements are unprotected wood of large cross-sectional dimensions. Columns must be at least 8 inches if supporting a floor load; joists and beams a minimum of 6 inches in width and 10 inches in depth. Type IV is not subdivided. The inherent fire-resistive nature of large diameter wood members is taken into account. Concealed spaces are not permitted (see Figure 4-4).

Type V: Wood Frame In type V construction (see Figure 4-5), the entire structure may be constructed of wood or any other approved material. Brick veneer may be applied, but the structural elements are wood frame. Type V is subdivided into protected and unprotected, again depending on the protection provided for the various structural elements.

Table 4-1 *Fire Resistance Rating Requirements for Building Elements (hours).*

Building Element	Type I		Type II		Type III		Type IV	Type V	
	A	B	A[d]	B	A[d]	B	HT	A[d]	B
Structural frame,[a] including columns, girders, trusses	3[b]	2[b]	1	0	1	0	HT	1	0
Bearing walls									
Exterior[f]	3	2	1	0	2	2	2	1	0
Interior	3[b]	2[b]	1	0	1	0	1/HT	1	0
Nonbearing walls and partitions									
Exterior	See Table 602 of the source document								
Interior[e]	See Section 602 of the source document.								
Floor construction, including supporting beams and joists	2	2	1	0	1	0	HT	1	0
Roof construction, including supporting beams and joists	1 1/2[c]	1[c]	1[c]	0[c]	1[c]	0	HT	1[c]	0

Note: For SI; 1 foot = 304.8mm.

[a] The structural frame shall be considered to be the columns and the girders, beams, trusses, and spandrels having direct connections to the columns and bracing members designed to carry gravity loads. The members of floor or roof panels that have no connection to the columns shall be considered secondary members and not a part of the structural frame.

[b] Roof supports: Fire-resistance ratings of structural frame and bearing walls are permitted to be reduced by 1 hour where supporting a roof only.

[c] 1. Except in Factory-Industrial (F-I), Hazardous (H), Mercantile (M), and Moderate Hazard Storage (S-1) occupancies, fire protection of structural members shall not be required, including protection of roof framing and decking where every part of the roof construction is 20 feet or more above any floor immediately below. Fire-retardant-treated wood members shall be allowed to be used for such unprotected members.

2. In all occupancies, heavy timber shall be allowed where a 1-hour or less fire-resistance rating is required.

3. In type I and type II construction, fire-retardant-treated wood shall be allowed in buildings not over two stories including girders and trusses as part of the roof construction.

[d] An approved automatic sprinkler system in accordance with Section 903.3.1 [of the source document] shall be allowed to be substituted for 1-hour fire-resistance-rated construction, provided such system is not otherwise required by other provisions of the code or used for an allowable area increase in accordance with Section 506.3 [of the source document] or an allowable height increase in accordance with Section 504.2 [of the source document]. The 1-hour substitution for the fire resistance of exterior walls shall not be permitted.

[e] For interior nonbearing partitions in type IV construction, also see Section 602.4.6 [of the source document].

[f] Not less than the fire-resistance rating based on fire separation distance (see table 602 [of the source document]).

Source: 2000 International Building Code®. Table 601, p. 89. *Copyright 2000, International Code Council, Inc., Falls Church, Virginia. 2000 International Building Code. Reprinted with permission of the author. All rights reserved.*

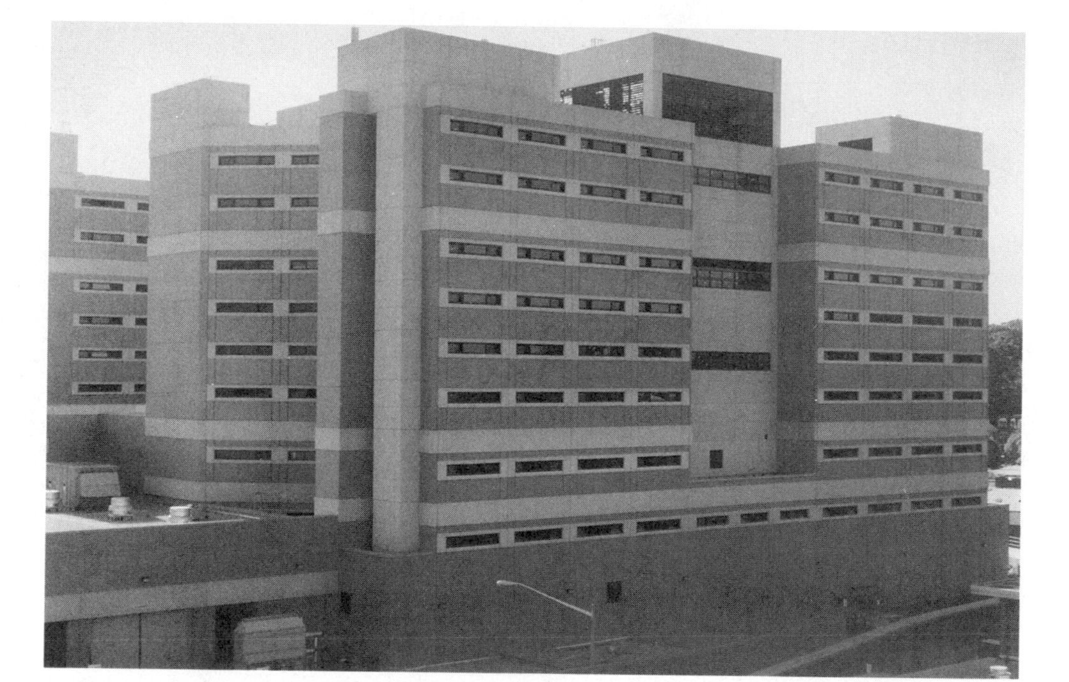

Figure 4-1 *Type I construction provides maximum fire resistance.*

Figure 4-2 *Type II construction is noncombustible but lacks the fire resistance of type I.*

Figure 4-3 *Type III "ordinary" construction is common in cities.*

Figure 4-4 *Type IV buildings feature noncombustible exterior walls and interior elements of large solid or laminated wooden members. Log structures are not type IV.*

Figure 4-5 *Type V buildings may be totally combustible.*

FIRE-RESISTANCE RATINGS

■ NOTE
Each of the model codes and NFPA 220 has a table containing the ratings (in hours) for the various structural elements.

Each of the model codes and NFPA 220 has a table containing the ratings (in hours) for the various structural elements. The tables list required ratings by building component type, depending on the construction classification of the building. The construction classifications used by the model codes and NFPA 220 do not exactly match type for type. Table 4-2 provides an approximate comparison.

Look at the table in your building code and compare the required rating for a column or structural frame member that supports more than one floor. Notice a big difference as you move from type I through type V Unprotected? How about for loadbearing exterior walls? We will see later that the code rewards fire-resistive design and penalizes the others.

HEIGHT AND AREA LIMITATIONS

If the children's story *The Three Little Pigs* had been written by a fire inspector, it might have been a bit more of an action thriller. Instead of huffing and puffing and blowing the straw and stick houses down, the big bad wolf could have set a fire and let one house burn down the other. The third pig, living in his masonry house equipped with a sprinkler system would still live happily ever after, perhaps making a fortune in real estate by buying up the vacant lots left behind by the fire.

Table 4-2 *Comparison of the construction classifications of the model building codes.*

IBC (2000)	IA		IB	IIA	IIB	IIIA	IIIB	IV	VA	VB
Table 601										
NFPA 5000 (draft)	I	I	II	II	II	III	III	IV	V	V
Table 7.2.2.2	443	332	222	111	000	211	200	2HH	111	000
BNBC (1999)	1A	1B	2A	2B	2C	3A	3B	4	5A	5B
Table 602										
SBC (1999)	I	II		IV	IV	V	V	III	VI	VI
Table 600				1Hr	U	1Hr	U		1Hr	U
UBC (1997)	I	II		II	II	III	III	IV	V	V
Table 6-A		FR		1Hr	N	1Hr	N		1Hr	N

He was a smart pig; after all, he built a type III building and his buddies built type V. He knew that all construction materials do not give the same performance.

■ **NOTE**
Physical characteristics of the materials, including the ability to resist ignition and maintain structural integrity when exposed to fire are what we have referred to as fire resistance.

There is an inherent risk associated with the type of materials used in the construction of buildings. Physical characteristics of the materials, including the ability to resist ignition and maintain structural integrity when exposed to fire, are what we have referred to as *fire resistance*. These characteristics are critical to the ability of the public to escape from the structure and, to limit fire spread to other structures, and for fire suppression forces to safely and effectively attack the fire.

The model building codes use height and area limits to establish equivalent risks for structures based on four critical elements:

1. Use group
2. Construction type
3. Building frontage or open space that provides exposure protection and fire department access
4. The presence of an automatic sprinkler system

■ **NOTE**
The model building codes use height and area limits to establish equivalent risks for structures.

All three building codes start with a base height and area figure taken from Tables 503 (*BNBC*), 5-B (*UBC*) or 500 (*SBC*).

The *IBC, BNBC, SBC,* and *UBC* start with a base height and area taken from a table, and then allow increases depending on the amount of open perimeter or building frontage and whether the building is fully sprinklered. However, the NFPA 5000 draft proposes a radically different method. Building height for sprinklered and nonsprinklered buildings are listed in separate tables. The concept of building area is abandoned and replaced with *compartment* areas. Under the *NFPA 5000* draft, buildings are separated into compartments. The size and rating of the envelope surrounding the compartment is dependent on the use of the space and sprinkler protection.

building height
according to the *IBC*, "the vertical distance from grade plan to the average height of the highest roof surface"

grade plane
an imaginary plane representing the finished ground around the exterior walls of a building

reference datum
a plane representing the elevation of the highest adjoining sidewalk or ground surface within a 5-foot horizontal distance of a building; used by the *UBC* in determining building height

grade
the average of finished ground level around a building or within a given distance from a building; used in determining building height

NOTE
In addition to height, the model building codes regulate the number of stories permitted.

Building Height

There are a couple of reasons that the model building codes require the vertical distance of a building to be measured. First is to determine whether the building meets the overall height limits set by the codes based on use group and type of construction. A second measurement is sometimes required by the code to determine whether the building must conform to the special requirements for high-rise buildings or requires the installation of standpipes.

The *2000 IBC* simplifies the matter of determining **building height** as being "the vertical distance from grade plane to the average height of the highest roof surface." Sounds easy, until you look at the definition of "grade plane" and just what is meant by "average height of the highest roof surface." To begin with, the **grade plane** is an imaginary plane representing the finished ground around the exterior walls of a building. If the ground level slopes away from the building, the grade plane is established using the lowest point within 6 feet of the exterior walls unless the lot line is closer.

The business about average height of the highest roof surface means top of a flat roof and the average height of the highest gable on a peaked roof. The *UBC* used the term **reference datum** to refer to the grade plane,[4] and the *SBC* simply used the term **grade**. the *NFPA 5000* draft measures building height from "grade to the average elevation of the highest roof surface."

In addition to height, the model codes regulate the number of stories permitted. The different codes use definitions of *basement*, *first story*, and *story above grade* to address portions of a building that are partially below grade and to determine whether a story is in fact a "basement" and not to be counted when determining whether the building meets the code limit on number of stories. This becomes especially important in modern multifamily residential buildings that are often of combustible construction and have large occupant loads (see Figure 4-6). The potential need for aerial and ground ladders and elevated streams gives new significance to how the area around a building is graded, and how well fire department access has been maintained.

In determining whether a building requires standpipes or must meet the high-rise requirements, a second measurement is taken from the lowest level of fire department vehicle access to the floor level of the highest occupied floor. This measurement is taken only on one side of the building—the lowest level that the fire department can drive equipment to. The measurement is made from the ground to the floor level of the highest floor that is occupied.

Figure 4-6 *This building is classified as four stories in height. (Courtesy of Carl Maurice.)*

■ **NOTE**

Building height may be modified from what is specified in the tables by the installation of a sprinkler system.

Height Modifications

Building height may be modified from what is specified in the tables by the installation of a sprinkler system. The *SBC* incorporates the sprinkler provisions into Table 500. The *BNBC* and the *UBC* allow building heights specified in Tables 503 and 5-B to be increased one story in most use groups if a building is fully sprinklered. *NFPA 5000* has different height, area, and separation tables for sprinklered and nonsprinklered buildings. The concept of regulating the size of fire areas within the building versus regulating the size of the *building* itself is one of the features that sets *NFPA 5000* apart from the *IBC* and its predecessors.

Area Modifications

The IBC defines building area as: *"The area included within surrounding exterior walls, or exterior walls and fire walls, exclusive of courts . . ."*[5] The other two codes have similar definitions. Area modifications could be described as changes to the *footprint* of the building, or the actual amount of ground space it covers.

All four codes allow the area to be increased based on how much of the building perimeter is open and accessible to the fire department. The *BNBC* includes an extra step, an area *decrease* for non-fire-resistant construction types depending on the height of the building. The taller the building, the greater the decrease (up to ten stories).[6]

The *UBC* area modifications are based on a formula that gives credit for the number of sides of the building that open on a yard or public way, 20 feet or more in width. The *BNBC* and *SBC* require a 30-foot public space or way and use formulas that credit the percentage of the perimeter that is open and exceeds 25 percent of the total. The *BNBC* also requires the installation of marked fire lanes within the public way in order to increase the building area.

Unlimited Area Buildings

The ultimate area modification is included in each of the model building codes and is called the *unlimited area building.* It is generally limited to business, factory, mercantile, and storage occupancies and is fully sprinklered. Noncombustible construction or setbacks on all sides of up to 60 feet, or both, may be required.

■ NOTE

The ultimate area modification is included in each of the model building codes and is called the *unlimited area building.*

EXPOSURE PROTECTION FOR BUILDINGS

The threat of fire spread to adjacent buildings and structures has been regulated by the model building codes in two ways. First, noncombustible and fire resistive construction are rewarded through the *allowable height and area* provisions of the code. Providing an open perimeter around the building enhances that reward. The second method is based on the distance from the building to the property line. The closer the building to its neighbors, the greater the exposure potential. The model building codes attempt to minimize fire spread to exposed buildings by increasing the required fire-resistance rating for exterior walls as the distance between buildings is reduced. Table 602 of the *IBC* lists ratings in hours based on distance, type of construction, and use group. Exterior wall protection and how we protect openings in exterior walls is discussed in Chapter 5.

Summary

Regulating the size of buildings based on the fire-resistive qualities of the construction materials used has been around for a long time. The National Board of Fire Underwriters' *National Building Code* contained provisions for height and area limitations and requirements for fire-resistive construction in 1905. These requirements provide for the safety of the occupants and establish exposure protection for adjoining structures. In establishing construction types, both combustibility and fire resistiveness of construction elements are considered.

Review Questions

1. ______________ is the resistance of a building to collapse or total involvement in fire.
2. When used in describing construction types *protected* means ______________
3. According to your building code, noncombustible unprotected construction is classified as type _____.
4. According to your building code, heavy timber construction is classified as type _____.
5. According to your building code, which types of construction must have noncombustible exterior walls? ______________
6. What four elements are used by the model building codes to establish limits on building height and area?

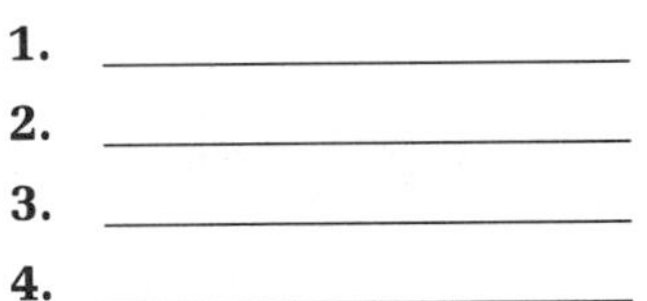

7. Building height is determined by measuring from grade or the reference datum to ______________ on a flat roof or ______________ on a pitched roof.
8. List two reasons why the model building codes provide area increases based on the amount of a building's perimeter that faces a street or open lot.
   1. ______________
   2. ______________

Discussion Question

1. The management of a condominium apartment complex has applied for a permit to operate a place of assembly. A seating plan for the second floor of the community swim club building showing 287 folding chairs with adequate aisles and exits is attached to the application. Before leaving the office for the inspection, you review the file and find the following information:

   Community Swim Club Building
   1234 Any Street
   2 Stories, nonsprinklered
   Gross floor area 7200 sq. ft./ floor

a. Is the building's use group classification appropriate for the intended use?

b. What steps must the management take to legally establish a place of assembly?

c. Is there anything about the building's size or construction that would preclude the establishment of a place of assembly in this building?

Fire-Resistive Construction Elements

Learning Objectives

Upon completion of this chapter, you should be able to:

- Describe the differences between specification codes and performance codes.
- Explain why a rated assembly must be maintained as a unit.
- Describe the origin of the standard time temperature curve, explain why it is still in use, and the significance of fire-resistance time values.
- List three nationally recognized testing laboratories involved in the testing of fire-resistance-rated assemblies.
- Explain why the labeling of certain rated assemblies is mandated by the codes.

performance code
a code that assigns an objective to be met and establishes criteria for determining compliance, e.g., requirements for fire assemblies rated in "hours" and building materials rated as "noncombustible" when tested to specific protocols

■ NOTE
The model codes are for the most part performance codes.

specification code
a code that specifies a type of construction or materials to be used

listed
equipment or materials included in a document prepared by an approved testing agency indicating that the equipment or materials were tested in accordance with an approved test protocol and found suitable for a specific use

assembly (rated assembly)
a building component such as a door, wall, damper, or ceiling composed of specific parts and tested and listed as a unit

Fire walls, fire partitions, fire barriers, fire separation assemblies—all are designed to resist the spread of fire, but are they the same things? Fireblocking and draftstopping are installed to prevent the free passage of flame and products of combustion within concealed spaces. Is one as good as the other? Are they interchangeable?

When considering fire-resistive construction, a point to remember is that the model codes are for the most part prescriptive documents. They "prescribe" or state a required level of protection. There are a number of ways to meet the prescriptive requirements. Rather than specify the exact construction of a component such as a fire wall, the code identifies the performance features of the component. Where a **specification code** might require all fire walls to be of a certain thickness of block, a performance code identifies the performance required (in hours). The designer can use any **listed** assembly that provides that fire-resistance rating. The term "performance based design," is a different animal altogether. Prescriptive requirements are abandoned and buildings are designed to achieve desired levels of fire resistance and life safety through the use of computer models and mathematical calculations.

All fire-resistive assemblies required by the code will have performance requirements prescribed. The building designer will have specified the **assembly** (by design number) to be used to attain the required rating, or will have used one of the approved equivalent fire-resistance methods from the code. The *IBC*, *UBC* and *SBC* contain specific provisions for structural protection. The important thing for the fire inspector to remember is that every rated assembly is composed of elements that are tested and listed as a whole. Whether a wall, a floor/ceiling assembly, or a door, the entire assembly is tested and listed together. If any portion of the assembly is left out or altered, the assembly is no longer the listed item, and more importantly, the assembly may not function as designed.

Examples of *prescriptive* or specification type code requirements, or requirements that are not performance based, are those that specify fireblocking and draftstopping materials:

> Fireblocking shall consist of 2-inch (51 mm) nominal lumber or two thicknesses of 1-inch (25 mm) nominal lumber with broken lap joints or one thickness of 0.719-inch (18.3 mm) wood structural panel with joints backed by 0.719-inch (18.3 mm) wood structural panel or one thickness of 0.75-inch (19 mm) particle board with joints backed by 0.75-inch (19 mm) particle board . . .. (2000 *IBC* 716.2.1)
>
> Draftstopping materials shall not be less than 0.5-inch (12.7 mm) gypsum board, 0.375-inch (9.5 mm) wood structural panel, 0.375-inch (9.5 mm) particleboard, or other approved materials adequately supported. (2000 *IBC* 716.3.1)

FIRE TESTS

Rated assemblies are subjected to tests specified in various standards including the National Fire Protection Association's NFPA 251, *Standard Methods of Tests of Fire Endurance of Building Construction and Materials*, the American Society

■ NOTE

Every rated assembly is composed of elements that are tested and listed as a whole.

standard time temperature curve (STTC) curve representing the standard reproducible test fire used since 1918 to measure the fire resistance of building materials

■ NOTE

Components are assembled in the laboratory and then subjected to fire conditions in accordance with the standard time temperature curve.

for Testing and Materials (ASTM) *ASTM E119*, Underwriters' Laboratories (UL) *UL 263*, and *UBC* Standard 7-1 through 7-7. NFPA 252 specifies *Standard Methods of Fire Tests of Door Assemblies.* Check your building code to find the standards referenced.

The components are assembled in the laboratory and then subjected to fire conditions in accordance with the **standard time temperature curve** (STTC) (see Figure 5-1). Acceptance criteria include:

- Ability to support a specified load
- Ability of the assembly to limit the transfer of heat to the unexposed side of the assembly
- Ability of the assembly to prevent the passage of heat or flame sufficient to ignite cotton waste
- Ability to prevent the passage of excess heat to steel members
- Ability to withstand exposure to a hose stream

It is important to remember that the values (in time) assigned to rated assemblies are *relative.* A 2-hour wall survived the test protocol twice as long as the 1-hour wall did. When exposed to actual fire conditions, it is possible that neither one might hold back a fire for an hour. The STTC used to regulate the intensity of the "fire conditions" that our rated assemblies are exposed to is based on a fuel load of 8,000 BTUs per pound, which is about what you get from burning wood and paper products. The plastic materials, which are so evident in our modern buildings, weigh in at about 16,000 BTUs per pound.[1]

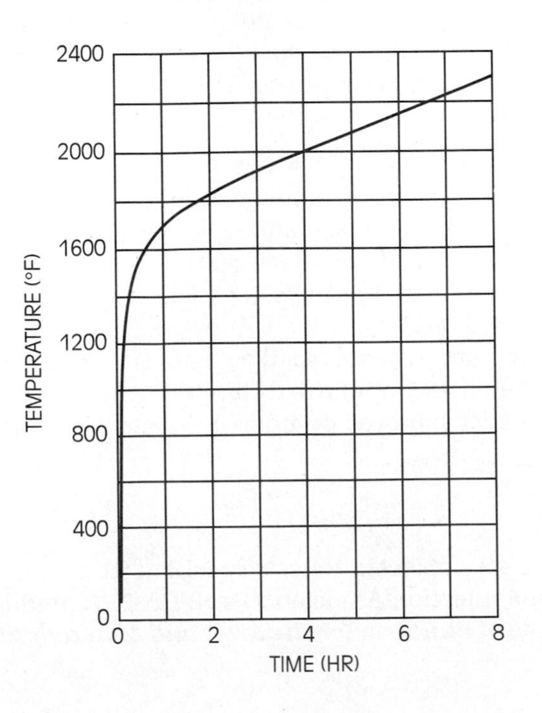

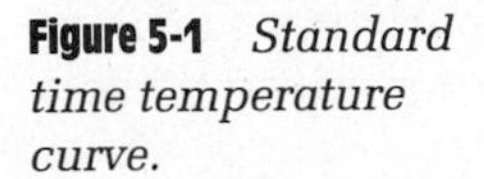

Figure 5-1 *Standard time temperature curve.*

■ NOTE
Values (in time) assigned to rated assemblies are *relative.*

■ NOTE
The "fire conditions" that our rated assemblies are exposed to are based on a fuel load of 8,000 BTUs per pound.

■ NOTE
Plastic materials, which are so evident in modern buildings, weigh in at about 16,000 BTUs per pound.

■ NOTE
When it comes to rated assemblies, the details are everything.

If you are wondering why we expose construction materials to test fires that do not necessarily mirror those found in modern buildings, you need to understand the origin of the standard time temperature curve. It has been around for a long time. The American Society for Testing and Materials was testing beams, girders, and columns using the STTC in 1918.[2] At that time you could expect about 8,000 BTUs per pound from a fully loaded building of "ordinary" construction. It was a reasonable measure of performance that approximated fire conditions of the time. It is still a reasonable *relative* measure of performance. A relative measure of performance is really all a rating is.

RATED ASSEMBLIES

The fire-resistance ratings of building materials are not designed to give the occupant or firefighter a gauge as to how long a building will be tenable during fire conditions. They are performance comparisons under fixed fire conditions that do not mirror fire conditions in most modern buildings.

Accept the test results for what they indicate. Under laboratory conditions, the specimen performed to a given standard when exposed to a fire that did not replicate those found under actual fire conditions in today's environment. Do not expect to be able to eat your lunch on one side of a 1-hour fire separation wall because the fire next door has only been burning for 15 minutes.

Perhaps the best-known source of listings is UL's *Fire Resistance Directory.* Other directories are produced by Warnock Hersey, the Gypsum Association, and others. Laboratories that perform tests of fire-resistive assemblies include the Factory Mutual System, Inchape Testing Laboratories, the Portland Cement Association, and Underwriters' Laboratories. Assemblies are specified on building plans by design number. By referring to the appropriate directory, a description and diagram of the assembly are available.

An important element that is often glossed over is the details crucial to the performance of the assembly. Fire separation walls specify nail or screw schedules with fasteners of specific lengths. The edges must be staggered between layers of drywall, may be required to be taped, and assemblies are sometimes product specific (see Figures 5-2 and 5-3). When it comes to rated assemblies, the details are everything.

In Chapter 4, we discussed the model code tables that list fire-resistance ratings for the five construction types. Conformance with these requirements can sometimes only be determined through the use of the listing documents and approved plans for the building.

Maintenance of Rated Assemblies

The life of a rated assembly is often short and full of danger. Plumbers, telephone and cable installers, and electricians seem driven to run their services through fire separation assemblies. New building owners just are not happy with the listed

GA FILE NO. WP 1716	2 HOUR FIRE	40 to 44 STC SOUND

GYPSUM WALLBOARD, METAL STUDS

Base layer 5/8" type X gypsum wallboard or veneer base applied parallel to each side of 3 1/2" 20 gage steel studs 24" o.c. and attached to studs and runner track with 1" Type S-12 drywall screws 12" o.c. **Face** layer 5/8" type X gypsum wallboard or veneer base applied parallel to each side of studs and attached to studs and runner track with 1 5/8" Type S-12 drywall screws 12" o.c. Stagger joints 24" o.c. each layer and side. Studs attached to each side of top and bottom track with 1/2" Type S-12 panhead screws, or welded.

Bracing: Lateral bracing shall be as described by assembly WP 1206. **Tested at 80 percent of design load. (LIMITED LOAD-BEARING)**

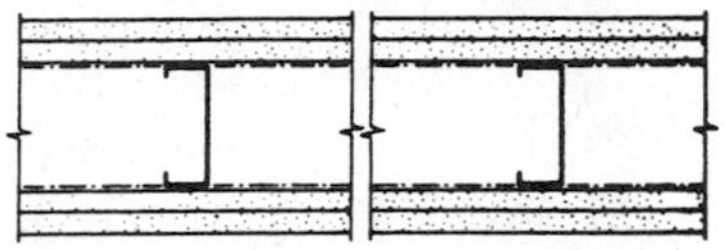

Thickness: 6"
Limiting Height: Subject to design
Approx. Weight: 10 psf
Fire Test: UL NC 505-6, 7-29-82, Design U425
Sound Test: See WP 1615

Figure 5-2 *Specifications for a 2-hour fire separation wall. (Courtesy of the Gypsum Association, Washington, DC.)*

GA FILE NO. FC 2030	2 HOUR FIRE	50 to 54 STC SOUND

STEEL JOISTS, CONCRETE SLAB, GYPSUM WALLBOARD

One layer 1/2" type X gypsum wallboard or veneer base applied at right angles to rigid furring channels with 1" Type S drywall screws 12" o.c. Wallboard end joints located midway between continuous channels and attached to additional pieces of channel 54" long with screws at 12" o.c. Furring channels 24" o.c. attached with 18 gage wire ties 48" o.c. to open web steel joists 24" o.c. supporting 3/8" rib metal lath or 9/16" deep 28 gage corrugated steel and 2 1/2" concrete slab measured from top of flute. Furring channels may be attached to 1 1/2" cold rolled carrying channels 48" o.c. suspended from joists by 8 gage wire hangers not over 48" o.c. (Two hour restrained and unrestrained.)

(See GA File No. BM 3310)

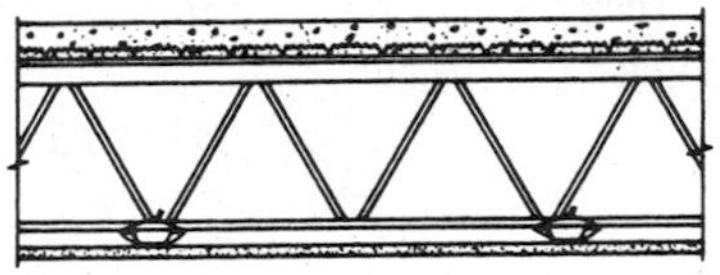

Approx. Ceiling Weight: 2 psf
Fire Test: UL R3501-28, 2-7-64, Design G514, ULC Design I511
Sound Test: NGC 4075, 3-25-69

Figure 5-3 *Specifications for a 2-hour floor/ceiling assembly. (Courtesy of the Gypsum Association, Washington, DC.)*

locks on the rated stairwell doors and frequently install nonlisted hardware. The lawyer has a residential mail slot installed in his rated office door. The employees at the auto body shop go one step further from propping open the rated exterior door—they also remove the self-closer. There is no level of fire protection that can be put into a building that a big enough hammer can not take out.

Ensuring the maintenance of fire-resistance rated assemblies is the responsibility of the fire inspector. Damage can be easily overlooked and often will be missed, *sometimes by several inspectors over the years.* The cry from the business owner that "This is the first time in five years that anybody mentioned it," doesn't make the potential hazard to the occupants and business owner any less. The rating must be restored.

The 2000 *International Fire Code* section 703.1 is very specific. "The required fire-resistance rating of fire-resistance rated construction (including walls, firestops, shaft enclosures, partitions, and floors) shall be maintained. Such elements shall be properly repaired, restored, or replaced when damaged, altered, breached, or penetrated." The other model fire codes have similar requirements. The 2000 *Uniform Fire Code* section 1111.1 adds "removed or improperly installed" to the mix.

label
a permanent identification affixed to a product by a manufacturer indicating the function and performance characteristics of the product, name of the manufacturer, and the name of the approved testing agency that tested a representative sample of the product

Labeling is an important safety element of the code. Fire doors, fire windows, fire shutters, and fire dampers all must bear the label of an approved testing agency. It shows the inspector that construction elements are listed for a particular application, and shows the contractor that his supplier is delivering the proper materials. Labeling costs money. Some manufacturers offer the same door with or without the label. The only difference is that one has a small embossed piece of metal attached.

The thought of saving a few dollars each on the hundreds of doors needed for a large apartment project by buying unlabeled units might appeal to the builder who is trying to maximize his profit. Same door or not, *they are unacceptable without the label.* The code is specific. The required labels are marks that will be checked by inspectors for the life of the building. Treat the label as a vital element of the assembly. It must be there.

■ NOTE
Fire doors, fire windows, fire shutters, and fire dampers all must bear the label of an approved testing agency.

■ NOTE
Treat the approved testing agency label as a vital element of the assembly; it must be there.

■ NOTE
In simplest terms, a fire wall is designed with the structural stability to be able to withstand the collapse of the structure on either side.

Fire Walls versus Fire Separation Walls

In order to understand the difference between the various structural components that are designed to prevent the spread of fire from one building to another, or one area within a building to another, go back to the building code definitions. The different codes have had their own terms for years. In formulating the first edition of the *International Building Code*, the members BOCA, ICBO, and SBCCI had to agree on common terms for the new building code. The development of NFPA 5000 is also requiring NFPA members to come up with terms for their new building code. One thing should be clear to every person who has reason to use any of the codes or referenced standards—*read the definitions.*

Fire walls, known as "area separation walls" by the *UBC*, are designed with the structural stability to be able to withstand collapse of the structure on either side and remain standing (see Figure 5-4). The wall does not have to be designed on the assumption that fires could start on both sides at the same time. The wall can rely on the structural stability of the buildings on either side, but must be designed with break-away bracing so that either side can collapse without causing the fire wall to collapse.[3] Fire walls may be parapeted, depending on the type of construction and combustibility of the roof and may be required to extend beyond the building's exterior walls.

Fire barriers, fire partitions, and fire separation walls are assemblies that sep-

Figure 5-4 *This fire wall will separate single-family town houses. It is being constructed in accordance with a listing from the Gypsum Association's design manual, and is rated for 2 hours.*

NOTE

Fire separation walls, fire barriers, and fire partitions are assemblies that separate and subdivide tenant spaces within a building; fire walls separate buildings.

arate and subdivide spaces within a building. The *IBC* mandates the use of a *fire barrier* to separate exits and exit passageways from the rest of the building, to separate mixed uses, individual fire areas, and to separate incidental use areas like boiler rooms. Fire barriers are rated assemblies that extend from the top of the floor/ceiling below to the underside of the floor/ceiling or roof deck above. Fire barriers extend through any concealed spaces such as ceilings, *even if they are rated.*

Fire partitions are used to separate dwelling units, guestrooms in hotels, tenant spaces within covered malls, and for corridor wall construction. With the exception of corridor walls, fire partitions are rated but are permitted to terminate at a rated ceiling.

Fire walls separate buildings. Fire barriers and partitions separate or subdivide areas within a single building. Generally walls that separate occupancies within strip shopping centers are not fire walls. There is a very good chance that the strip shopping center where you buy your groceries, pick up your dry cleaning, and rent videos is one big building with multiple tenants. The tenant separation walls may be fire barriers depending on the use groups involved and the type of construction used. More often than not, they are nonrated partition walls used solely for security purposes. This is an important issue during a fire incident. Fire officers who assume that tenant separation walls extend to the roof deck, or have a fire-resistance rating, may be in for a surprise. The total area within the four exterior walls may not have a single rated element within the entire shopping center.

Another significant issue involves the handling of regulated materials using the fire codes. Maximum allowable quantities/Exempt amounts for materials are *per building*, not *per tenant space*.

Exterior Opening Protectives

As you recall from Chapter 4, type of construction is one of the factors that determines the required fire-resistance rating for the exterior walls of the building. The other factors are use group and fire separation distance. The 2000 *IBC* defines fire separation distance as: "The distance measured from the building face to the closest interior lot line, to the centerline of the street or public way, or to an imaginary line between two buildings on the property. The distance shall be measured at right angles from the lot line.[4]"

opening protective
a rated assembly such as a door or window that provides a protected opening in a rated wall or partition; there are interior as well as exterior opening protectives

Depending on these factors, the exterior of the building, including openings such as doors and windows, may have to be provided with specific fire-resistance ratings (see Figure 5-5). The number and size of openings may be limited, or in some cases prohibited. Ratings may be reduced if the building is fully sprinklered. The intent of the codes is to limit the spread of fire from one building to those adjacent. The rating required for the **opening protectives** is not necessarily the same as for the walls themselves. An exterior wall that requires a 2-hour fire-resistance rating under the *International Building Code* would require 1½-hour-rated opening protectives (714.2). In all cases, exterior openings that require protection must be provided with a minimum of ¾-hour rated assemblies.

Figure 5-5 *Fire shutters provide exposure protection. (Courtesy of Lionel Duckwitz.)*

■ **NOTE**
The rating required for the opening protectives is not necessarily the same as for the walls themselves.

■ **NOTE**
The most abused fire-resistance-rated assembly is the fire door, because it has moving parts and comes in contact with humans more than any other assembly.

Fire Door Assemblies

The most abused fire-resistance-rated assembly is the fire door, simply because it has moving parts and comes in contact with humans more than any other assembly. Fire doors come in a multitude of shapes and sizes. These include single side swinging, side swinging doors in pairs, dutch doors, and horizontal and vertical rolling fire doors.

The *assembly* (see Figures 5-6 and 5-7) includes the door, frame, and all of the hardware including the hinges, latches and locks, closing mechanism, and any special features such as fusible links. These assemblies are tested and listed as a whole. They must be installed and maintained as a whole. The installation of locks, hold-open devices, or other special features after the fact, void the listing and violate the integrity of the rated assembly, unless the additional items are themselves listed for installation *on that door* and are installed in accordance with the manufacturer's specifications. As with exterior opening protectives, the re-

Figure 5-6 *This fire door held back heat and smoke, enabling firefighters to escape a fire in their station. (Courtesy of Duane Perry.)*

Figure 5-7 *The latching mechanism is a key element of a fire door. (Courtesy of Duane Perry.)*

quired rating for fire doors in a rated wall or partition may be less than the wall itself. Each of the building codes has a table that lists the minimum opening protection for the various rated wall assemblies.

Structural Members

The ratings of columns, girders, trusses, beams, and other structural members are provided by a listed method such as encasement in concrete, a gypsum assembly, or with a spray-on encasement product (see Figures 5-8 and 5-9). The maintenance of this protection can be difficult if the areas are subject to pedestrian or vehicular traffic, industrial activity, or tradesmen who decide that the assembly is "in the way" of their cable or pipe installation.

■ NOTE

Column impact protection should be treated by the fire inspector as an integral part of the assembly.

Impact protection is required by the building codes and is provided for columns by the installation of corner guards or a substantial jacket of metal or other noncombustible material. Column impact protection should be treated by the fire inspector as an integral part of the assembly. He should order any damage to the guards or metal jacket to be repaired. If the impact protection method is ineffective, the owner should be ordered to provide an adequate method. Care should be taken to ensure that any new method of protection does not damage the rating of the structural member.

Figure 5-8 *Spray-on coatings provide fire-resistance for steel building components. (Courtesy of Duane Perry.)*

Figure 5-9 *Sheetrock has been added to these columns for aesthetics and to protect the fireproofing from impact damage. (Courtesy of Ron Berry.)*

■ NOTE
Fireblocking and draftstopping are often taken for granted until a room and contents fire burns off the roof of an apartment building.

Fireblocking and Draftstopping

Fireblocking and draftstopping are often taken for granted until a room and contents fire burns off the roof of an apartment building or the crew that was opening up windows on the floor above the fire reports that they are now operating a hose line. *Draftstopping* is designed to prevent fire spread within large concealed spaces such as attics and floor assemblies that use web trusses or have suspended ceilings.[5]

Fireblocking is used to prevent the spread of fire within small concealed passages such as floors, walls, and stairs.[6] The maintenance of these features is often complicated because access to concealed spaces is limited. Often the only persons who enter are tradespeople who run a new half-inch electrical cable through a 6-inch square hole, or building occupants who decide that void spaces make great storage spaces.

The requirement for the draftstopping of attic spaces has changed over the years. The 1970 *BOCA Basic Building Code* required firestopping in the attic every 3,000 square feet unless the roof and attic were noncombustible.[7] By 1981, draftstopping was required in line with tenant separations unless the building and attic was sprinklered.[8]

Fireblocking is installed at ceiling and floor levels of combustible construction, as well as at connections between horizontal and vertical concealed spaces. Openings around vents, ducts, chimneys, and fireplaces are also fireblocked.

Fire Dampers

Fire dampers are usually required to be installed wherever air distribution systems penetrate a rated assembly. They must be listed for the particular application and labeled. Like fire doors, the fire damper rating and the rating of the penetrated assembly are not always the same. The *IBC, NBC, UBC,* and *SBC* all permit 1½-hour fire dampers in 2-hour assemblies. (See Figure 5-10).

Access for maintenance and inspection of fire dampers is required by the building codes. If you can not get to it to inspect it, then a technician can not get to it to maintain it. Lack of access is a serious problem that must be dealt with to ensure the integrity of the rated assembly.

Other Assemblies

Other fire-resistance-rated assemblies, including fire windows, fire shutters, wired glass panels, and other opening protectives (see Figure 5-11) all must be maintained in accordance with their listings by the owner of the building or structure. Fire inspections should include a check of testing and maintenance records, as well as a visual check of the assemblies themselves. A thorough inspection requires a look above that false ceiling every so often and a good look at any new utility work. Snoop around a little; good inspectors are nosy about the building they are inspecting.

Figure 5-10 *Fire dampers are installed where ducts penetrate rated assemblies. This damper failed because it was installed using unapproved screws that blocked the path of one wing of the damper, locking it open.*

Figure 5-11 *Openings in roof/ceiling assemblies must be equipped with listed opening protectives. (Courtesy of Howard Bailey.)*

Summary

The building code requirements for fire-resistant construction features are generally perspective code rather than performance code requirements. They prescribe the required performance in hours. Requirements for structural elements vary according to the type of construction, and can be found in the model codes in table form.

Assemblies are tested under fire conditions according to the test protocols using the standard time temperature curve. NFPA 251 for building materials and NFPA 252 for fire doors are examples of the test standards. Assemblies are tested and listed in directories by Underwriters' Laboratories, Warnock Hersey, the Portland Cement Association, and other approved testing agencies. The rating (in hours) should not be considered an absolute indication of how long the assembly will withstand actual fire conditions, because the test protocol does not replicate actual fire conditions in modern buildings.

Assemblies must be listed for the application, and some assemblies such as fire doors, fire dampers, fire shutters, and fire windows must bear the label of an approved testing agency.

Review Questions

1. For the most part, the model building codes are considered to be ______________ type codes.
2. A code provision that requires all fire walls to be constructed of a certain thickness of masonry would be considered a __________ code provision.
3. Rated assemblies and components are tested under fire conditions prescribed by the ______________.
4. Name three organizations that list fire-resistance-rated assemblies.
   1. ______________
   2. ______________
   3. ______________
5. List two fire-resistance-rated assemblies that must bear a label.
   1. ______________
   2. ______________
6. What is the primary difference between a fire wall and a fire barrier?
7. ______________ is designed to prevent fire spread within large concealed spaces such as attics.
8. Wherever air distribution systems penetrate a rated assembly, a ______________ must be installed.
9. Fire-resistance ratings for listed assemblies are expressed in terms of __________.
10. Is the fire-resistance rating of an assembly a good indicator of its actual performance under fire conditions? __________.

Discussion Question

1. During your inspection of a plumbing supply warehouse, the owner is proud to inform you that business has been so good that he has expanded. He has purchased the building next door and doubled the size of his operation, expanding his 10,000-square-foot warehouse into 20,000 square feet by making two 10-foot openings in the block wall that separated the two nonsprinklered occupancies.

   a. What are some of the code provisions that address this expansion?

   b. What should the owner have done prior to expanding?

   c. What actions would you take in this case?

Installation of Fire Protection Systems

Learning Objectives

Upon completion of this chapter, you should be able to:

- List the four classes of fire and give examples of each.
- Identify fire suppression system agents that are compatible with the different classes of fire and describe the method of extinguishment provided by each.
- List and describe the four basic types of sprinkler system.
- Describe the differences between the NFPA 13, 13D, and 13R sprinkler standards and the application of each.
- Describe the function of standpipe pressure-reducing valves and the potential hazards posed by improper installation.
- Describe the purpose for the development of the UL-300 standard for restaurant cooking equipment.

fire protection system equipment and devices designed to detect a fire, sound an alarm and (or) make notification, control or remove smoke and hot gases, and control or extinguish the fire

Fire protection systems is a broad term that includes built-in extinguishing systems such as sprinklers and halon, detection and alarm devices, smoke control and removal equipment, fixed protection for cooking operations and industrial equipment, and even portable devices such as fire extinguishers. Each is designed for a specific hazard and is installed in accordance with the code to address that hazard. If the hazard changes, the fire protection system must be reevaluated to determine its potential effectiveness given the new condition.

FIRE-EXTINGUISHING SYSTEMS

NOTE
If the hazard changes, the fire protection system must be reevaluated to determine its effectiveness given the new condition.

Systems that automatically control or extinguish fires have become commonplace in our world. The cheeseburger you choked down at lunch today was hopefully cooked on a grill that was installed as required by code, under an approved exhaust hood and duct system. If there had been a fire on the grill while your order was cooking, it would probably have activated a wet chemical extinguishing system, suppressing the fire and shutting off all of the protected cooking appliances. You might not have eaten that cheeseburger.

Although you might not have noticed, the restaurant was also probably protected by a fire protection system. Most people do not pay attention to sprinkler heads—they have become part of the scenery people expect in a modern building.

Classification of Fires and Extinguishing Agents

Fires are classified according to the material or fuel involved. Fire-extinguishing agents are rated according to this classification system to ensure that the agent is compatible with the hazard. This is important for two reasons. First, the physical properties of the agent must match the hazards presented. Discharging a 2½-gallon water extinguisher onto a burning circuit breaker panel would definitely be an example of the use of an incompatible agent. Because water conducts electricity, the user would be at considerable risk of electrocution, and the fire would not be extinguished.

NOTE
The physical properties of a fire-extinguishing agent must match the hazards presented.

Our second consideration regarding agent compatibility is between agents themselves. Some extinguishing agents, notably different dry chemical compounds can react with one another, diminishing their effectiveness. Table 6-1 includes examples of the four classes of fire, with appropriate extinguishing agents.

Extinguishing Fires

In order to understand the various methods of fire extinguishment, an explanation of fire is in order. This may seem elementary, but fire investigators spend extended periods on the witness stand establishing their credibility as expert witnesses. They are routinely asked for a definition of *fire*. Some of the definitions that have been published in texts over the years go to the point of stating that fire is only *sometimes* accompanied by the evolution of light and heat *in varying degrees*.

Table 6-1 *Classification of fires and extinguishing agents.*

Fire Class	Description	Examples	Extinguishing Agents
A	Common combustibles	Wood, paper, cloth, plastic	Water, dry chemical, foam, some halon
B	Flammable liquids and gases	Gasoline, oils, grease, LPG	CO_2, dry chemical, halon, foam
C	Energized electrical equipment	Energized Class A material such as a household appliance	CO_2, dry chemical, halon
D	Combustible metals	Magnesium, sodium, potassium	Dry powder

fire
rapid oxidation accompanied by the evolution of heat and light

Since we are not on the stand, we will settle for the definition that I remember from my days in recruit firefighter training. **Fire** *is the rapid oxidation of a material, accompanied by the evolution of light and heat.* Fire is a chemical chain reaction. To put out the fire, you must stop the reaction. The older texts on fire prevention and firefighting explained the reaction by using a triangle (see Figure 6-1). The three sides are *heat*, *fuel*, and *oxygen*.

By removing any one of the sides, the reaction is halted and the fire is extinguished. Remove the heat by cooling the fire with a hose stream; deprive it of oxygen by blanketing the fire with carbon dioxide; or deplete its source of fuel say, by shutting down the leaking gas line, and the fire will be extinguished. All three methods have proven effective and are the basis for many modern fire extinguishing systems.

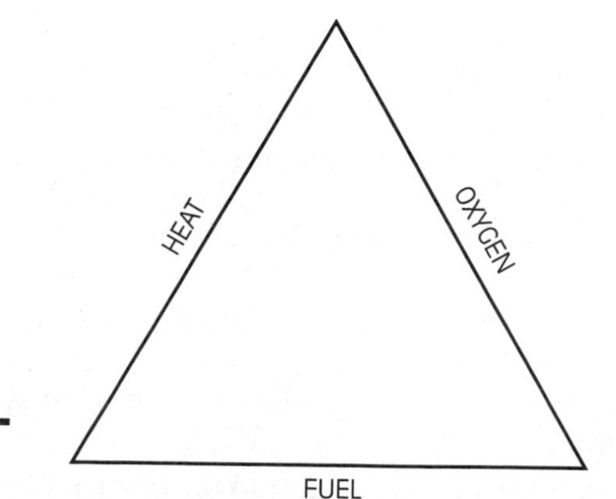

Figure 6-1 *The fire triangle.*

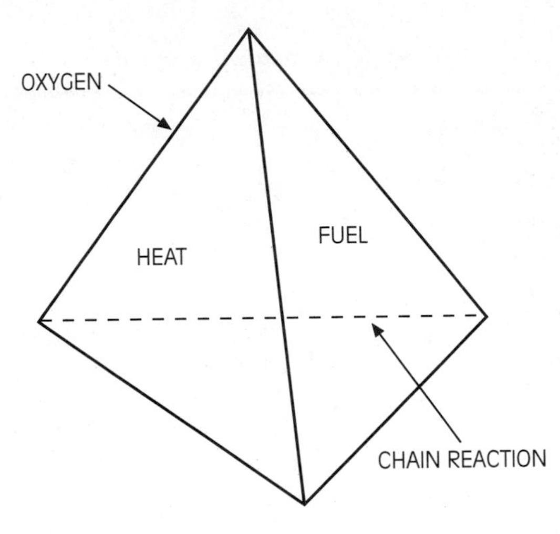

Figure 6-2 *The fire tetrahedron adds an extra element to the fire triangle.*

■ **NOTE**
The old fire triangle is also the basis for many of the provisions of our modern fire prevention codes.

The old fire triangle is also the basis for many of the provisions of our modern fire prevention codes. Regulations for the storage and handling of materials and for the use of open flames and heating appliances are all geared toward keeping the three sides of the triangle effectively segregated from one another.

Modern fire research led to the update of the fire triangle. An extra element, the molecular chain reaction, was added to create the fire tetrahedron (see Figure 6-2). With the new element came another method of fire extinguishment, interruption of the molecular chain reaction. Certain extinguishing agents such as dry chemical and halon extinguish fires by this method.

SPRINKLER SYSTEMS

■ **NOTE**
By far the most prevalent, economical, and efficient fire protection system in use today is the sprinkler system.

By far the most prevalent, economical, and efficient fire protection system in use today is the sprinkler system. The advantages over other systems are significant. What other system uses an extinguishing agent that is cheap, plentiful, nontoxic, and usually supplied to the premises by the municipality? Consider this:

- Ordinary combustibles produce about 8,000 BTUs per pound when fully involved in fire. One pound of water is capable of absorbing nearly 1,000 BTUs (970.3 to be exact).[1] One gallon of water (8.33 pounds) can absorb over 8,000 BTUs.
- Sprinkler systems activate quickly, activating an alarm and discharging water only on the fire area. Water damage is considerably less than when fires are manually extinguished by firefighters, minimizing business interruption.
- Sprinkler systems are in service around the clock and are fully automatic.

History and Development

The forerunners of the automatic sprinkler system were perforated pipe and open sprinkler systems installed in mill properties from about 1850 until 1880. Neither system was automatic and relied on supply valves being manually operated in the event of fire. Application at the seat of the fire was generally insufficient and water damage was extensive. Since all the discharge orifices in both systems were open, the entire area covered by the system was equally wetted.

In 1878, the first truly automatic sprinklers were installed. Technological improvements quickly followed, led by Frederick Grinnell of the General Fire Extinguisher Company.[2] The company, now bearing Grinnell's name, is still prominent in the sprinkler industry today.

NFPA 13, *Standard for the Installation of Sprinkler Systems*, was first printed in 1896. Since that time it has been updated and expanded. Technological advances have necessitated the breakdown of the sprinkler standard into multiple standards.

Sprinkler Standards

The demand for lower cost sprinkler systems to protect one- and two-family dwellings and smaller multifamily buildings led to the development of NFPA 13D and NFPA 13R (see following list). Inspection, maintenance, and testing provisions for all water-based fire protection systems were recently compiled into a single standard, NFPA 25. The *IBC, BNBC* and *SBC* adopt the NFPA sprinkler standards by reference, the *UBC* adopts NFPA 13 and 13R by transcription as UBC Standards 9-1 and 9-3 respectively.

Sprinkler Standards Referenced by the Model Codes

NFPA 13 UBC 9-1	*Installation of Sprinkler Systems*
NFPA 13D	*Installation of Sprinkler Systems in One- and Two-Family Dwellings and Manufactured Homes*
NFPA 13R UBC 9-3	*Installation of Sprinkler Systems in Residential Occupancies Up to and Including Four Stories in Height*
NFPA 25	*Inspection, Testing and Maintenance of Water-Based Fire Protection Systems*

The 1999 edition of NFPA 13, *Standard for the Installation of Sprinkler Systems*, was expanded to include sprinkler requirements for hazards such as rack and pile storage and special hazards such as rolled paper, plastic commodities, baled cotton, and rubber tires that were previously contained in other NFPA standards.

Additional sprinkler requirements are also found within NFPA standards that pertain to specific processes and occupancies. NFPA 30, *Flammable and*

■ NOTE
The notion that all sprinklered buildings are equally protected, regardless of the contents or system design, is a dangerous one.

Combustible Liquids Code, and NFPA 30B, *Manufacture and Storage of Aerosol Products*, are examples. Each of the standards provides a significant and different level of protection to address a specific hazard or hazards. The notion that all sprinklered buildings are equally protected, regardless of the contents or system design, is a dangerous one, embraced unfortunately, by too many in the fire and building regulation communities.

NFPA 13 Systems NFPA 13 systems are by far those most commonly found installed in buildings in the United States. NFPA 13, *Standard for the Installation of Sprinklers*, is the direct descendent of the original sprinkler standard developed by NFPA in 1896. As we learned in Chapter 1, the existence of nine radically different standards for sprinkler pipe size within 100 miles of Boston was a primary reason for the establishment of the National Fire Protection Association. NFPA 13 has been actively maintained by the organization and used throughout the world for over 100 years.

■ NOTE
The NFPA 13 sprinkler system is a property protection system that has had remarkable success at protecting people by quickly suppressing and often extinguishing fires in their incipient stages.

The NFPA 13 sprinkler system is a property protection system that has had remarkable success at protecting people by quickly suppressing and often extinguishing fires in their incipient stages. Heat from the fire rises to the ceiling where sprinklers are installed and act as combination heat detection and fire suppression devices. Once the operating elements of the sprinkler head (generally a solder link or frangible bulb) reach their operating temperature, they fail and water is discharged onto the fire through the open orifice.

Standard automatic sprinklers are designed to operate at specific temperatures. Ordinary temperature-rated heads, which operate at between 135° and 170°, are required to be installed throughout buildings unless the area is subject to high heat conditions. Boiler rooms, skylights, attic spaces, and areas above certain machinery and equipment may all require the use of sprinkler heads that have higher operating temperatures. Table 3-2.5.1 in NFPA 13 (1999) contains seven temperature ranges from Ordinary to Ultra High.

thermal lag
the difference between the operating temperature of a fire detection device such as a sprinkler head and the actual air temperature when the device activates

The standard attempts to minimize the amount of time that it takes for the sprinkler system to discharge water on a fire by requiring that Ordinary temperature heads be installed. This is important if fire spread is to be checked and early notification accomplished. In the design of sprinkler heads, **thermal lag** becomes the enemy.

A sprinkler head that is rated to operate at 165° does not operate the minute that the surrounding air reaches that temperature. The *detection device itself* must become heated to that temperature before it will operate.[3] In the case of a fire involving readily combustible material or flammable liquids, the surrounding temperature will have continued to rise and may be significantly higher by the time the head operates and water is discharged.

■ NOTE
Sprinkler head spacing and discharge per square foot, known as density, are dependent on occupancy classification.

Thermal lag is minimized by designing sprinkler heads and other detection devices with less mass and greater surface area. This concept played an important role in the design of sprinklers for residential applications and is discussed with the other sprinkler standards.

Table 6-2 *Occupancy classifications from NFPA 13.*

Classification	Examples
Light Hazard	Churches, hospitals, schools, offices, theaters, residences
Ordinary Group 1	Parking garages, canneries, laundries, restaurant service areas, electronic plants
Ordinary Group 2	Dry cleaners, horse stables, machine shops, post offices, print shops, library stack rooms
Extra Hazard Group 1	Aircraft hangars, saw mills, plywood manufacturing
Extra Hazard Group 2	Flammable liquid spraying, plastic processing, varnish and paint dipping.

Source: NFPA 13, 1999, p. 168, Appendix A.

NFPA 13 (see Table 6-2) uses an occupancy type system of hazard classification to determine the operational requirements of sprinkler systems, *which is not to be confused with the use group classification from the building codes.* Sprinkler head spacing and discharge per square foot, known as **density**, are dependent upon occupancy classification.

density
refers to sprinkler density calculated by gallons per minute discharge divided by the square footage covered

Using the occupancy classifications as a means to design sprinkler systems to meet the hazard that exists in a particular occupancy, building owners get an adequate system to protect their interest and that of their insurer, at a minimum cost. NFPA 13 systems provide excellent property protection by requiring that the entire building, including combustible void spaces such as attics be sprinklered.

NOTE
By quickly suppressing fire, flashover is prevented and occupants are able to escape from the building.

A quick look at the examples in Table 6-2 may bring to mind some of the occupancies within your city or county where the original owner has sold or leased the building to a business with a very different hazard classification. Ordinary Group 1 buildings are limited to storage heights of 8 feet, Ordinary Group 2 are limited to 12 feet. Determining that the hazard protected by the system has not changed is a key element in the inspections process, which we discuss in rather lurid detail in Chapter 10.

Specific requirements for the design, installation, and acceptance testing of sprinkler systems are all contained within NFPA 13. Requirements for the maintenance and routine testing of sprinkler systems are contained in NFPA 25, *Standard for the Inspection, Testing and Maintenance of Water-Based Fire Protection Systems.* Sprinkler requirements for special hazards such as aerosol products, flammable and combustible liquids, nitrate film, and others are included in other NFPA standards. NFPA 13 Chapter 7 directs the user to the specific standards.

NOTE
NFPA 13R systems do not, however, provide the same level of property protection as afforded by NFPA 13 systems.

NFPA 13R Systems NFPA 13R systems are designed to protect residential occupancies up to four stories in height. The systems enhance life safety by using listed residential sprinkler heads, which are designed to minimize thermal lag and respond

flashover
point at which the contents of a room or space becomes heated, simultaneously ignites, and the entire room or space becomes involved in fire

quickly in the event of fire. By quickly suppressing the fire, **flashover** is prevented and occupants are able to escape from the building.

NFPA 13R systems *do not*, however, provide the same level of property protection as afforded by NFPA 13 systems. Combustible voids, as well as closets not exceeding 24 square feet and bathrooms not exceeding 55 square feet are *not* required to be sprinklered. The total building area exempted from sprinkler protection has been estimated as high as 67 percent.[4] The potential property damage resulting from fires that "burn around, over and under" the sprinkler system is significant. The possibility of a "total loss" in a 13R sprinklered building is very real. Hotels, motels, boarding houses, and apartment buildings are typically protected with NFPA 13R systems.

A trash fire on the balcony of the apartment building in Figure 6-3 quickly threatened the attic. Protected with an NFPA 13R sprinkler system, neither the bal-

Figure 6-3 *Balconies and attics are unprotected in these sprinklered apartment buildings. (Courtesy of Carl Maurice.)*

conies nor the attic spaces were sprinklered. Only quick action by the fire department saved the building.

NOTE
All sprinkler systems are not equal. They are designed for specific hazards and to perform specific functions.

NFPA 13D Systems NFPA 13D systems are designed to protect one- and two-family dwellings and mobile homes. Like NFPA 13R systems, they are primarily designed to protect the occupants by preventing flashover from occurring and giving the occupants time to escape. They have similar limitations to NFPA 13R systems, in that bathrooms and closets as well as garages, attached porches, carports, and some foyers are not required to be sprinklered. All sprinkler systems are not equal. They are designed for specific hazards and to perform specific functions.

Types of Sprinkler Systems

wet pipe sprinkler system
an automatic sprinkler system in which the supply valves are open and the system is charged with water under supply pressure at all times

fire line
dedicated underground supply piping for a sprinkler or standpipe system

limited area sprinkler system
an automatic sprinkler system that is limited to a single fire area and consists of not more than twenty sprinklers

There are four basic types of sprinkler systems: wet pipe, dry pipe, preaction and deluge. What differentiates them is the method in which water is supplied to the system. Each has specific applications and system-specific requirements from the sprinkler standards and in a few cases from the building codes themselves.

Wet Pipe Sprinkler Systems Wet pipe sprinkler systems are by far the most commonly found systems in most jurisdictions. Water enters the sprinkler system from a dedicated supply or **fire line**, unless the system is a **limited area sprinkler system** or a system designed to protect one- and two-family dwellings and mobile homes. The supply valve is left open and water at street pressure is always on the system. For this reason, they are the quickest at getting water on the fire and are the simplest to maintain.

Wet pipe systems are installed where indoor temperatures can be maintained at 40°F. Below that temperature, there is the danger of freezing pipes. If you recall learning that water freezes at 32° F, your memory serves you well. If the outside temperature is below freezing and the interior temperature is less than forty, the steel sprinkler piping, which rapidly conducts heat and rapidly loses it, will drop below freezing. The frozen area may be isolated and near an opening or uninsulated portion of the building. It may be a small area, but it could be enough to put the whole system out of service.

NOTE
Wet pipe systems are installed where indoor temperatures can be maintained at 40°F, because below that temperature, there is the danger of freezing.

Antifreeze systems are sometimes used where freezing is expected, but these systems are usually limited to 40 gallons or less. Each year, antifreeze systems must be drained into containers and the antifreeze solution must be replenished to restore the specific gravity required to prevent freezing. This process is costly and labor intensive. Larger systems subject to freezing are normally designed as dry pipe.

The wet pipe system's advantage of getting water to the fire area quickly is also its weakness in areas where water damage is of great concern. Mechanical damage from machinery such as forklifts or from vandals can lead to the soaking of valuable equipment, artifacts, merchandise, or records. The public's perception that all the sprinkler heads within a system discharge at once has caused undo concern on the part of many. Concern for items that may not be replaceable is legitimate and is easily addressed.

dry pipe sprinkler system
an automatic sprinkler system that features dry piping maintained under constant air pressure and a dry pipe valve in which water is held back by the pressure differential between the system air pressure and water supply pressure; used where there is a danger of freezing

dry pipe valve
sprinkler water supply valve designed to permit a moderate amount of air pressure above the valve to hold back a much greater water pressure from the incoming supply

differential
ratio of air pressure to water pressure that is necessary to balance a dry pipe valve, maintaining it in the closed position

Dry Pipe Sprinkler Systems Dry pipe sprinkler systems are installed in warehouses, parking garages, factories, and other buildings where there is a danger of freezing. They may also be installed in unheated attics or portions of buildings where providing heat is impractical. A **dry pipe valve** (see Figure 6-4) is installed at the sprinkler riser which keeps water out of the system piping until a fire activates a sprinkler head or heads. Dry pipe valves are designed so that a moderate amount of air pressure in the system above the valve (see Figure 6-5) is capable of holding back a much greater water pressure. The ratio of air pressure to water pressure at which the valve will open or trip is called the **differential**. Dry pipe systems are equipped with two pressure gauges. One gauge measures the incoming water pressure below the dry pipe valve, and the other measures the air pressure on the system above the valve.

Air pressure for the system piping is usually supplied with a small compressor. The amount of air pressure required is dependent on the designed differential of the particular type of valve and the water supply pressure at the

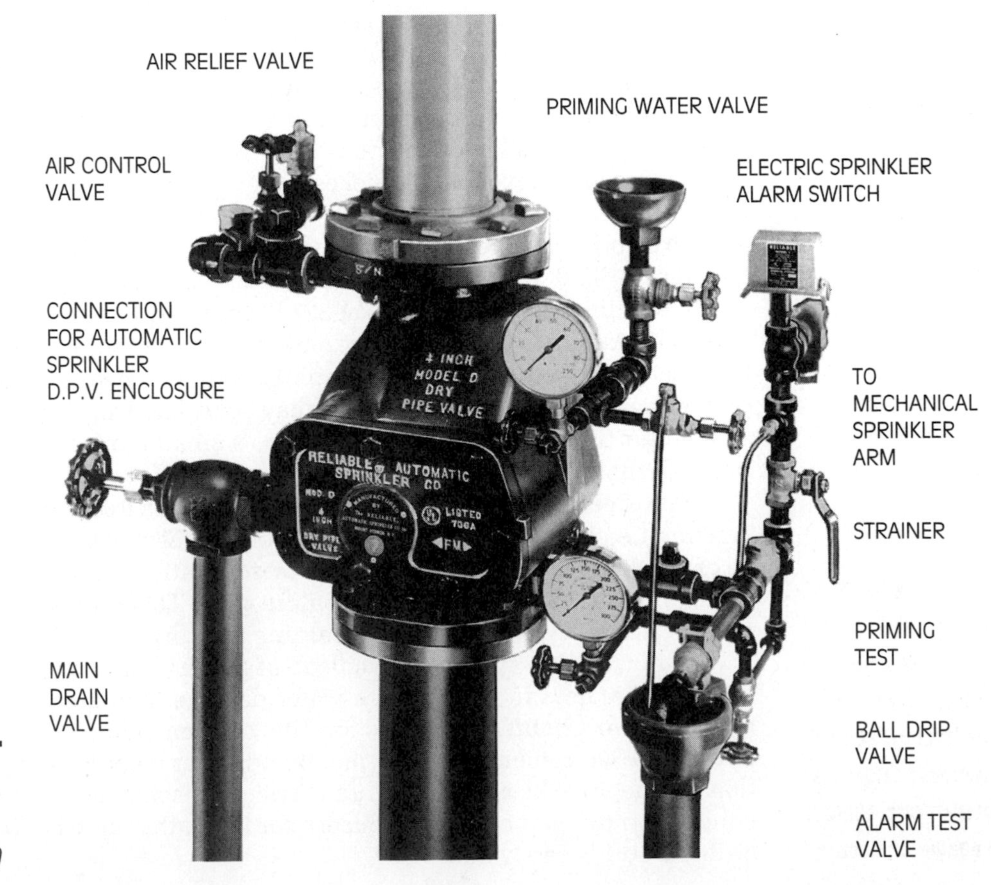

Figure 6-4 *Dry pipe valve. (Courtesy of Reliable Automatic Sprinkler Company.)*

sprinkler riser. Differentials of 5 or 6 pounds of incoming water pressure to 1 pound of air pressure are common with traditional designs. Low differential valves, which operate with ratios of 1.0 and 1.2 pounds of water pressure to 1 pound of air pressure, have the advantage of up to a 70 percent reduction in the time required for water to flow from the system.[5]

The fusing of a sprinkler head causes air pressure to drop within the system to a point where the dry pipe valve is tripped and latches open. Water then enters the system, displaces the air, and finally is discharged from the heads that have opened over the fire. The delay between the point when the dry pipe valve trips and when water is discharged is the fundamental weakness of these systems. This delay, which is effectively managed in most cases, makes the systems inappropriate for certain high hazard occupancies.[6]

In order to minimize this time lag, the sprinkler standards require dry pipe systems of more than 500 gallon capacity to be equipped with quick-opening devices,[7] unless water can be delivered to the most remote point of the system within 60 seconds. **Accelerators** and **exhausters** are designed to speed up the process. The *BNBC* requires that all dry pipe systems deliver water to the inspector's test pipe, which is the most remote point in the system, in not over 60 seconds.[8]

■ NOTE
The delay between the point when the dry pipe valve trips and when water is discharged onto the fire is the fundamental weakness of these systems.

accelerator
a quick opening device that permits system air pressure to enter the dry pipe valve below the one-way clapper valve, unbalancing the differential and causing the valve to trip more quickly

exhauster
a quick opening device used in dry pipe sprinkler systems that uses an auxiliary valve to discharge system air pressure to the atmosphere

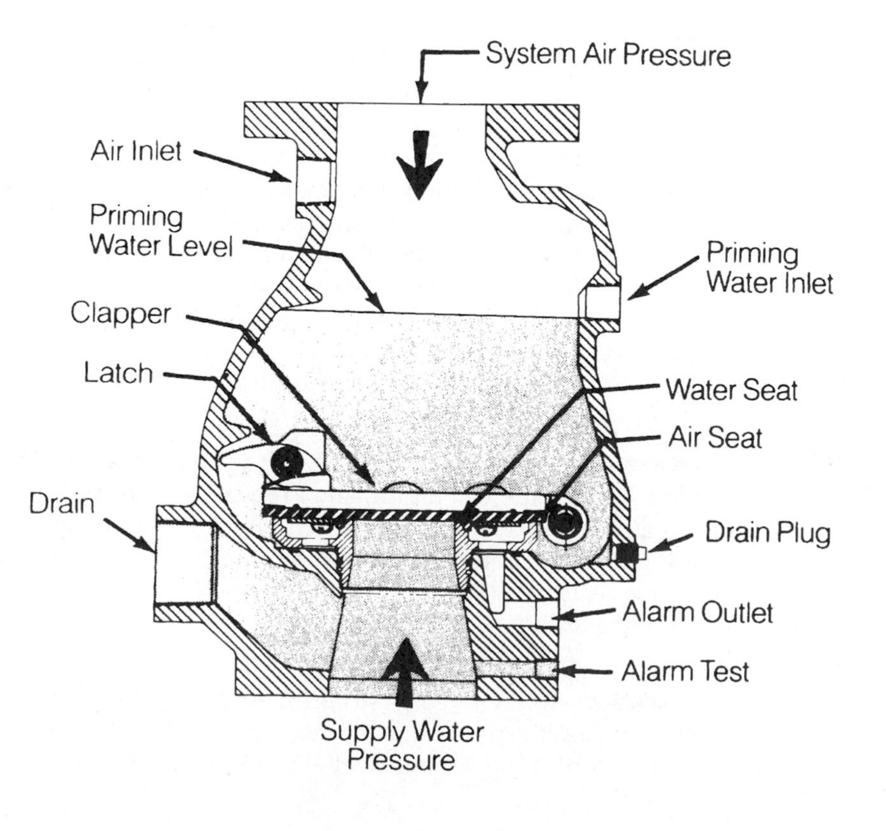

Figure 6-5 *Dry pipe valves hold back water with air pressure. (Courtesy of Reliable Automatic Sprinkler Company.)*

The sprinkler riser and dry pipe valve must be installed within a heated enclosure and care must be taken to ensure that the system does not have low sections of piping where water can become trapped and not drained. Equal pressure readings on both gauges at the dry pipe valve indicate that the valve has tripped and the system is now wet. It will operate as a wet system but is subject to freezing. Inspection and maintenance is discussed in Chapter 10.

preaction sprinkler system
an automatic sprinkler system that features dry piping, standard fusible heads, and a water supply control valve that is activated by fire detection devices; used to reduce the possibility of water damage from accidental breakage or discharge

Preaction Sprinkler Systems Preaction sprinkler systems are installed in properties where potential water damage from broken piping or sprinkler heads is of particular concern. Preaction systems are dry systems in which the water supply valve is opened on a signal from detection devices such as heat or smoke detectors. The potential for accidental water damage is minimized because no water is in the system unless a detection device has been activated or the valve is manually opened.

The sprinkler heads in a preaction system are traditional closed heads that must be fused by heat to open, discharging water only over the fire. Early notification also minimizes water damage associated with suppression of the fire because an alarm is transmitted upon activation of the detection device, which precedes the fusing of sprinkler heads and activation of a sprinkler flow switch.

deluge sprinkler system
an automatic sprinkler system that features open heads, dry piping, and a deluge valve that controls the supply of water, designed to wet the entire area upon activation

deluge valve
sprinkler water supply valve that is activated automatically or manually

Deluge Sprinkler Systems Deluge sprinkler systems are installed in extra hazard occupancies where there is the possibility of a flash fire or fire growth so rapid that the response of a standard sprinkler system is too slow. Facilities with large quantities of flammable liquids or materials that pose a deflagration hazard are protected with these systems. The activation of the system is designed to apply water over the entire area covered by the system, rapidly and simultaneously. Deluge systems are dry systems that use open sprinkler heads and a **deluge valve**. The deluge valve operates much like the supply valve in a preaction system. When activated by a fire detection device, the deluge valve is automatically opened and water is discharged from all of the heads. Manual activation is also provided. Aircraft hangers, flammable liquid tank vehicle loading racks, and industrial facilities that process flammable or explosive materials are among those protected by these systems.

■ **NOTE**

The witnessing of acceptance tests by code officials is time wisely invested.

Acceptance Tests Acceptance tests are required for all sprinkler systems and include a hydrostatic test and flush of the system piping, system operational tests, and main drain flow test. Hydrostatic testing of underground fire service mains (see Figure 6-6) should be conducted prior to backfilling. The witnessing of acceptance tests by code officials is time wisely invested. Table 6-3 summarizes acceptance testing requirements from NFPA 13.

early suppression fast response sprinklers (ESFR)
a type of fast response sprinkler designed to suppress fires in high challenge fire hazards through the application of increased flow densities

ESFR Sprinklers ESFR, or **early suppression fast response** sprinklers, were designed to protect palletized and rack storage up to 35 feet in some cases, with sprinklers mounted only at the ceiling level. ESFR systems use sprinkler heads with typical

Figure 6-6 *This fire service main must be hydrostatically tested before backfilling.*

Table 6-3 *Acceptance tests from NFPA 13.*

Test	What/How	Why
Underground piping flush	Flow at 10 ft/sec through 4″ min pipe or at max flow rate	Flush rocks and debris before connection to the system
Hydrostatic test of piping	200 psi for 2 hours	Ensure system integrity under operating pressures
Dry pipe leakage test	40 psi for 24 hours	Ensure piping is capable of holding air under pressure
System operations tests	Trip test of dry pipe valve, test of deluge valve, pressure-regulating devices and test of water flow devices and alarms	Ensure that all valves, quick opening devices, and alarms function
Main drain flow test	Open main drain valve, record static and residual pressure	Ensure all supply valves are open,used to compare with future tests to verify water supply

Source: NFPA 13, 1999, p. 150 and 151.

■ NOTE
ESFR systems use sprinkler heads with typical flow rates of up to 125 gallons per minute per head, or about five times as much as a standard sprinkler.

flow rates of up to 125 gallons per minute per head, or about five times as much as a standard sprinkler. ESFR sprinklers are designed to extinguish fires. Standard sprinkler systems have always been designed to confine fires by prewetting combustibles that surround the fire.[9]

The technology is very attractive for storage occupancies. Storage racks can be reconfigured without sprinkler modifications, and storage heights are increased. As with any new technology, there are bugs to be worked out. Duct work and lighting fixtures that might be installed by tenants after the sprinkler system is installed and approved can seriously compromise its effectiveness. Like any fire protection system, inspection to determine that the operational effectiveness of the system has not been reduced is of critical importance.

STANDPIPES

Standpipe requirements are contained within the model building codes and in the referenced standpipe standard. The *IBC, NBC* and *SBC* adopt NFPA 14, *Standpipe and Hose Systems,* by reference and the *UBC* adopts it by transcription as UBC Standard 9-2, *Standpipe Systems.*

Standpipe systems are classified according to their intended use, by the building occupants, fire department, or both. Table 6-4 summarizes standpipe supply requirements by type.

Standpipes are generally required in high-rise buildings, covered mall buildings, buildings over three stories, large assembly occupancies, and adjacent to stages in theaters. Each of the model building codes has slightly different requirements. Generally, standpipe connections are required on each floor at every exit stairway, and on each side of a horizontal exit.

Standpipes must be capable of supplying 500 gpm at 65 psi (pounds per square inch) at the topmost outlet for the first standpipe, and 250 gpm for addi-

Table 6-4 *Standpipe classes from NFPA 14.*

Class	Minimum Supply	Intended User
I	500 gpm for 30 minutes 250 gpm for additional standpipes Must maintain 65 psi residual @ 500 gpm	Fire department
II	100 gpm for 30 minutes Must maintain 65 psi residual @ 100 gpm	Building occupants
III	Same as Class I 2½″ to 1½″ reducers are installed to facilitate the use of the system by building occupants	Fire department and building occupants

Source: NFPA 14, 1996 edition.

tional standpipes in the building. The 65 psi predates the acceptance of the automatic nozzle by the fire service, and then did not account for the friction loss in the 2½-inch hose line.

Buildings under construction are generally required to have at least one standpipe or temporary standpipe, capable of flowing 500 gpm within one floor of the top of the building. These requirements are found within the respective fire prevention codes and are further discussed in Chapter 10, but are important to note here because they are not in all four building codes. You would not want to have to explain to the chief why the standpipe was not there.

Acceptance Tests for Standpipes

standpipe pressure-regulating device
a valve permanently attached to the standpipe discharge and designed to reduce flow pressure to a predetermined level by restricting the orifice size

Acceptance tests for standpipes are similar to those for sprinkler systems and include the all-important flush, hydrostatic test, and flow test from the most remote discharge. A note here about **standpipe pressure-regulating devices.** As a fire pump pushes water up the standpipe, it must overcome the head pressure that the water column in the standpipe exerts at the base of the riser. It is .434 psi per foot or about 5 psi per floor. In order to pump water to the second floor and supply 100 psi of pressure at the standpipe, you would add an extra 5 psi or so. In order to get that same 100 psi at the standpipe on the twentieth floor you would have to add about 100 extra psi (20 × 5 psi). It gives you 100 psi on the twentieth floor, but 195 psi on the second floor (100 + (19 × 5)).

Standpipes are limited to 275 feet. Buildings that exceed 275 feet must have standpipes zoned to address this pressure problem, limiting the number of floors a standpipe can cover. Pressure-regulating devices are designed to control pressures within a single zone. The devices attach to the 2½-inch standpipe connections and restrict the opening, thereby limiting the available pressure. Two points crucial to the efficiency of the standpipe system as well as to the safety of the firefighters:

- The devices are designed for installation in a particular system and on a particular floor.
- They are preset at the factory and cannot be adjusted on the fire scene.

■ NOTE
If the buildings in your jurisdiction have not been inspected to determine that all pressure-regulating devices are installed on appropriate floors and will discharge correct flow and pressure, you are waiting for a tragedy to occur.

In our example of the twenty-story building, a pressure-regulating device on the fourth floor might reduce the opening of the orifice on the standpipe enough to decrease the available pressure by over 100 psi. What happens if that valve is mistakenly installed on the twentieth floor? A crew operating off that standpipe connection might advance a hose line into a fire area and find that they cannot get water.

The problem starkly came to the attention of the fire service when three Philadelphia firefighters were killed at the One Meridian Plaza fire.[10] If the buildings in your jurisdiction have not been inspected to determine that all pressure-regulating devices are installed on appropriate floors and will discharge correct flow and pressure, you are waiting for a tragedy to occur.

Fire department connections are required for all water-based fire extinguishing systems and standpipe systems with the exception of limited area sprinkler systems of less than twenty heads.

OTHER EXTINGUISHING SYSTEMS

Wet and Dry Chemical Extinguishing Systems

Wet and dry chemical extinguishing systems are used in various applications including range hood and duct fire protection, paint spray booth protection, and even at unattended self-service motor vehicle fueling sites (see Figure 6-7). Each has an NFPA standard that addresses the design, installation, and maintenance of the

Figure 6-7 *Dry chemical nozzles are located above the island and at ground level. (Courtesy of Brad Cochrane.)*

systems (NFPA 17A and 17 respectively). They use four significant mechanisms of extinguishing fires: smothering, cooling, radiation shielding, and chain breaking.[11]

The development of the UL-300 Standard for restaurant cooking areas by Underwriters Laboratories has generated considerable discussion within the fire community. The standard was adopted by UL in November of 1994 as a result of significant changes in the hazards posed in commercial cooking processes. First, cooking equipment has become more efficient. Heating elements are positioned on all four sides and on the bottom of fryers, not just on the bottom. Fryers are larger and much better insulated.

■ NOTE
In a commercial cooking operation, if the cooking equipment has not been upgraded and the frying medium is the same, then the hazard is the same and the system should be considered adequate.

Where 50 pounds of oil was standard 20 years ago, 80 pounds is not unusual today, and in order to satisfy the health-conscious public, vegetable oils have replaced animal fats. Vegetable oils burn at higher temperatures than animal fats. If the quantity of overheated frying medium has doubled, then the heat absorption capability of the extinguishing system must also. Three to five times as much wet chemical agent as was previously required is required to meet the new UL-300 standard.[12]

Just as important for the fire inspector to realize is that this is not a retroactive requirement. Systems that have been previously approved (see Figure 6-8) have not been declassified by UL. The key issue here is whether the hazard has changed. If the cooking equipment has not been upgraded and the frying medium

Figure 6-8 *Nozzles must protect all cooking appliances except fully enclosed ovens and auxiliary cooking appliances that do not produce grease-laden vapors.*

is the same, then the hazard is the same and the system should be considered adequate. Acceptance tests for wet and dry chemical extinguishing systems are prescribed by their respective standards and should always be witnessed by the code official.

Halogenated and Clean Agent Fire Extinguishing Systems

Halon, carbon dioxide (CO_2), and other extinguishing agents that do not leave a residue are the agent of choice for certain high-value commodities and equipment. During the late 1970s and early 1980s, halon was an acceptable topic of conversation at computer systems administrator's cocktail parties. Many seemed convinced that the only type of sprinkler system was the deluge type, which in the event of fire would destroy their computer systems and lead to the financial ruin of their corporations. Many computer rooms were equipped with halon systems in lieu of sprinklers.

halogenated extinguishing agent (halon)
a clean extinguishing agent composed of carbon and one or more elements from the halogen series (fluorine, chlorine, bromine, and iodine), which leaves no residue

Halogenated extinguishing agents, or halon, are chemical compounds that contain carbon and one or more elements from the halogen series (fluorine, chlorine, bromine, and iodine). Most of the halogenated agents are toxic and unsuitable as extinguishing agents. Two compounds, Halon 1211 and Halon 1301, are effective extinguishing agents and are considered nontoxic.

Halon 1301 is used in total flooding systems, which discharge quantities sufficient to create up to a 10 percent concentration in the areas protected. Halogenated agents suppress fires by interrupting the chemical chain reaction, and do not extinguish deep-seated fire in class A materials.

Under the Clean Air Act, the United States government banned the production and importation of halons 1211 and 1301 effective January 1, 1994, in compliance with the Montreal Protocol On Substances That Deplete the Ozone Layer. The ban raised questions regarding the continued viability of existing halogenated systems as well as halon fire extinguishers. The United States Environmental Protection Agency offers the following information at its web site:[13]

■ NOTE
Existing halon fire protection systems can continue in service since there are no laws that prohibit halon emissions, although EPA, NFPA, and others all discourage the discharge testing of halon.

> Existing halon fire protection systems can continue in service. There are no laws which prohibit halon emissions, although EPA, NFPA and others all discourage the discharge testing of halon.
>
> Recycled halon can be purchased to recharge existing systems. The United States owns over 40 percent of the world's supply of halon 1301. The Halon Recycling Corporation or HRC, a nonprofit clearing house assists buyers and sellers.

In light of the government ban on halon manufacture and importation, other clean extinguishing agents such as Great Lakes Corporation's FM200, Dupont's FE-13 and Ansul's INERGEN have been developed. Fire protection for areas where the prevention of accidental water damage is a concern has also been rethought. The

■ NOTE
The major drawback of CO_2 systems is that carbon dioxide in concentrations of more than about 9 percent will render persons unconscious almost immediately.

use of preaction sprinkler systems, which nearly eliminate the possibility of accidental discharge, has increased significantly.

Carbon Dioxide Extinguishing Systems

Carbon dioxide is one of the most plentiful compounds on earth. It is a by-product of combustion and fermentation. We exhale carbon dioxide with every breath. It is cheap, leaves no residue, and is a highly effective extinguishing agent for class B and C fires. The major drawback to CO_2 systems is that carbon dioxide in concentrations of more than about 9 percent will render persons unconscious almost immediately,[14] so is not suited for the total flooding of computer rooms and other occupied areas. NFPA 12 *Carbon Dioxide Extinguishing Systems* contains installation, maintenance, and testing procedures.

Fixed Foam Systems and Water Spray Systems

Fixed foam and water spray systems are generally installed to protect hazards outside of buildings. Foam systems protect flammable liquid storage facilities and water spray systems are generally for exposure protection. Both have associated NFPA standards and specific acceptance tests that should always be witnessed by code officials, preferably by the end user, the fire department.

Fire extinguishers are covered in depth in Chapter 10. They are nonetheless, a vital element of a buildings's fire protection, which compliment built-in systems. Fire extinguishers are required to supplement some extinguishing systems such as range hood suppression. As with all other fire-extinguishing appliances and systems, the extinguishing agent must be compatible with the hazard and the other agents in use.

FIRE ALARM SYSTEMS

Fire alarm systems are required in certain buildings so that occupants receive prompt notification to evacuate, or in the case of institutional occupancies, to take appropriate action. Fire alarm components are required in specific locations in all buildings. Smoke detectors are required in elevator lobbies to ensure that elevators do not return to the fire floor. Air handling units over 2,000 cubic feet per minute (cfm) must automatically shut down upon activation of a duct smoke detector installed upstream from the air handler. Smoke detectors are required in the sleeping areas of all residential occupancies and must be interconnected within individual dwelling units.

The design, installation, and maintenance of fire alarm systems are regulated by NFPA 72. NFPA 72, 1996, is the consolidation of the previous editions of

NFPA 71, 72 72E, 72G, 72H, and NFPA 74 into one document entitled the *National Fire Alarm Code*.

Unlike most humans, fire alarm systems are quite good at doing several things at once. Fire alarm systems are comprised of a series of devices and circuits normally linked by a control panel. **Detection devices** sense the presence of heat or smoke, sprinkler flow, or manual activation by a building occupant, and signal the control panel. **Signaling devices** display audio and visual signals to the building occupants that there is a potential or actual fire in the building. Fire alarm systems also perform auxiliary functions such as elevator recall, automatic actuation of smoke removal or stairwell pressurization fans, and shutdown of certain air handlers. The system may also notify the fire department or a central station monitoring company.

detection device
a device connected to a fire alarm system having a sensor that responds to physical stimulus such as heat or smoke

signaling device
a notification appliance, being an alarm system component such as a bell, horn, speaker, light, or text display that provides audible, visible, or tactile output

Fire alarms range in size from the single station, battery-powered smoke detector in an existing one-story home to a complex system in a high-rise building. Each has detection devices, signaling devices, and a control center. In the five dollar battery-powered unit, they are all housed in a 6-inch plastic box. In a high-rise building, the circuitry may go for miles and cost many thousands of dollars.

Most complex modern systems are computerized. Fire alarm technicians do a significant amount of work within the software program during installation. When the flow switches are found to be annunciating in the fire control room as duct smoke detectors, technicians simply go into the program and make the correction with a few key strokes. Before the advent of computerized systems, a guy with a screw driver and pair of pliers had to move some wires to accomplish such a change—and that's the potential problem.

NFPA 72 requires acceptance testing of 100 percent of the system, and reacceptance testing any time components are added, deleted, or modified. This includes changes to software.[15] Additionally, reacceptance tests must include 10 percent of the initiating devices *not affected by the modification*. This requirement is necessitated by the complexity of modern systems and the sheer number of devices and functions involved.

Schedules for routine testing and maintenance of systems as well as requirements for completion documents and for records of all tests are contained in NFPA 72. A review of these documents should be a routine part of the inspection of the building. Do not wait until after the fire when the system has malfunctioned to ask if the system was being maintained.

EMERGENCY ALARM SYSTEMS

The 2000 *IBC* and *IFC* include requirements for emergency alarm systems for high-hazard occupancies. These are *not* fire alarms. They are systems that detect toxic and highly toxic gases, flammable gases, and refrigerants, and have manual pull stations that transmit a signal to a constantly attended location. They are people protection systems.

SMOKE CONTROL SYSTEMS AND SMOKE AND HEAT VENTS

■ NOTE

NFPA 72 requires acceptance testing of 100 percent of the system, and reacceptance testing any time components are added, deleted, or modified, including changes to software.

Systems that control the movement of smoke or provide for the rapid exhaust of smoke are required in atriums, covered malls, high-piled combustible storage facilities, underground structures, and large theaters. High-rise buildings and malls are required to provide manual controls for all air handling equipment so that firefighters can use the equipment for ventilation of the structure.

Smoke and heat vents are required by the model building codes in large factory and storage buildings where the length of exit access travel is long and over stages due to the large quantity of combustible sets. Curtain boards, which extend from the ceiling a minimum of 6 feet (but not within 8 feet of the floor), are installed to retard the lateral movement of smoke and gases. Smoke and heat vents must operate automatically and may be required to be provided with a means for manual activation by the fire department.

Summary

Fire protection systems are hazard-specific by design. The method of extinguishment as well as the capabilities of the system are designed to address a specific application. If that hazard changes, the system must be reevaluated.

All sprinklered buildings are not equally protected. NFPA 13, 13R, and 13D provide different levels of protection against different hazards. NFPA 13 provides protection to all areas of the building where combustibles (including building materials) may be present. In suppressing and perhaps extinguishing fires, they provide an excellent level of protection to the occupants of the building.

Residential sprinkler systems (NFPA 13R and 13D) provide a decreased level of property protection but optimize survivability of a fire by persons within the fire area by the quick response of the system. When installed in conjunction with smoke detectors, residential sprinkler systems provide an extremely high level of safety for building occupants from fires that originate within their dwelling unit.

The notion that fire protection construction features such as fire walls can be totally replaced by sprinkler systems is a dangerous one. Fire walls between buildings provide property protection for each building owner. They are protected against accidents or the irresponsible actions of their neighbors by an assembly that requires no maintenance and is not easily breached. For a code official to trade that protection away on behalf of all the future owners of the property in return for a system that can be turned off or fail is questionable. Especially if the sprinkler system is designed for life safety, not property protection.

Thorough testing of the systems, both during the approval or acceptance phase and for the life of the systems, is extremely important. The interests of the building owner as well as the contractor are best served by a formal program of 100 percent acceptance testing of all systems, witnessed by the fire official.

Review Questions

1. List the four classes of fire.
   1. ______
   2. ______
   3. ______
   4. ______
2. Fire is the rapid ______ of a material, accompanied by the evolution of heat and light.
3. What are the three legs of the fire triangle?
   1. ______
   2. ______
   3. ______
4. What are the four sides of the fire tetrahedron?
   1. ______
   2. ______
   3. ______
   4. ______

5. The difference between the operating temperature of a sprinkler head and the actual air temperature when the head operates is called ________________.

6. Residential sprinkler systems installed in accordance with the NFPA 13R or UBC 9-3 standards are designed to prevent ________________ and allow occupants time to escape.

7. List three areas in which sprinkler heads may be omitted in buildings equipped with sprinkler systems designed under the NFPA 13R or UBC 9-3 standards.
   1. ________________
   2. ________________
   3. ________________

8. List the four basic types of sprinkler systems.
   1. ________________
   2. ________________
   3. ________________
   4. ________________

9. ESFR sprinkler system stands for ________________.

10. List two changes in the commercial cooking process that led to the development of the UL-300 Standard.
    1. ________________
    2. ________________

Discussion Questions

1. A restaurant owner calls you to question the action of a fire equipment service company that attached a noncompliance tag to his range hood suppression system. They informed him that his system did not comply with the UL-300 standard and must be replaced. He informs you that since he fries very little, he is using the original cooking equipment from 1986, when the system was approved.
   a. Must he install a new extinguishing system?
   b. If not, in what circumstances would a new system be required?

2. You receive a call from the building inspector. He advises that a contractor has connected an underground fire main without flushing the system. The building is equipped with a standpipe and sprinkler system. The contractor proposes flushing the system with a 2½-inch hose line from the top standpipe outlet and flowing off the roof.
   a. Is this an acceptable method of flushing the system?
   b. How should the system have been flushed?

Means of Egress

Learning Objectives

Upon completion of this chapter, you should be able to:

- Identify and define the three components of a means of egress.
- Define the terms *common path of travel, dead end, design occupant load, horizontal exit,* and *travel distance.*
- Describe what is meant by the terms *gross floor area* and *net floor area* in the *BNBC* and *SBC*, and *floor area* in the *UBC.*
- Describe the six steps used in this chapter to design or assess the means of egress from a building or space.
- Calculate the required means of egress from a multipurpose space.
- Calculate the maximum occupant load for an assembly occupancy based on the exits provided.

■ NOTE

All of the concepts incorporated within the code are important, but if one had to be singled out for its role as the key element in many of our nation's disastrous fires it would be *means of egress*.

EGRESS AND THE CODES

The modern codes system employs a series of complimentary fire safety concepts. Our building regulations are a bit of a balancing act designed to give us reasonably safe structures. All of the concepts incorporated within the code are important, but if one had to be singled out for its role as the key element in many of our nation's disastrous fires it would be *means of egress*. Often, other deficiencies such as combustible interior finish or the improper use of open flame devices are compounded by inadequate means of egress. In the end, the ability of the occupants to quickly and efficiently exit the building is often the difference between life and death.

The Iroquois Theater fire, which occurred in Chicago in 1903, resulted in the deaths of more than six hundred people. It was later described as a "trifling fire"[1] that was quickly extinguished by the fire department, and was confined to the stage area and seats immediately in front of the proscenium. In the aftermath, bodies were reportedly piled three and four high in the aisles leading from the gallery.

Designed for 1,602 seats, over 1,770 were in place and nearly three hundred people were standing in the aisles behind the last row. The large volume of combustible theatrical sets and the lack of an appropriate fire separation between the stage and the audience, combined with overcrowding and exits that were obstructed or locked to create the worst theater disaster in our nation's history.

City ordinances that required skylights over the stage, an asbestos proscenium curtain designed to provide a separation between the stage and audience in the event of fire, flame-resistant scenery, and automatic sprinklers had simply not been enforced.[2] In the end, the "trifling" fire led to the deaths of more than six hundred people because they just could not get out of the theater quickly enough.

In order to ensure that occupants are able to safely leave a building in any situation (not just fire emergencies), the building codes regulate construction features of the egress facilities and the occupant load. The codes:

- Establish a design occupant load based on the use of the space.
- Determine the number and capacity of exits based on the design occupant load.
- Establish a maximum travel distance from any point within the space to an exit.
- Require that multiple exits from a space be sufficiently remote from one another in case any one exit is compromised.
- Establish minimum ratings for corridors and exit enclosures.
- Establish criteria for direction of door swing, size, and methods of locking and latching.
- Require adequate lighting of all portions of the means of egress.

■ NOTE

There are three distinct components of a means of egress: the exit access, the exit, and the exit discharge.

exit access
that portion of a means of egress from any point in a building to an exit

exit
that portion of a means of egress that is separated from all other parts of a building by rated assemblies and provides a protected path to the exit discharge

exit discharge
that portion of a means of egress between the exit and the public way

travel distance
the length of the path a building occupant must travel before reaching an exterior door or an enclosed exit stairway, exit passageway, or horizontal exit; the total length of the exit access

horizontal exit
an exit from one building to another on approximately the same level; or a passage through or around a rated wall or partition that affords protection from fire or smoke coming from the area from which escape is made

ELEMENTS OF THE MEANS OF EGRESS

There are three distinct components of a means of egress: the **exit access**, the **exit**, and the **exit discharge**. All model code groups include the definitions within the code. Such a breakdown may seem trivial until we consider code provisions for maximum **travel distance**. The code is not regulating the distance from every point in a building to the door at the sidewalk. It is regulating the distance from the corner office in a high-rise building to the enclosed, fire-rated stair tower, or the distance in the shopping mall from the bookstore to the fire rated exit enclosure that runs behind the stores. In a high-rise building, the *exit access* begins at any point on the floor and terminates inside the rated stairwell. The *exit* begins inside the door of the rated stairway and terminates outside the stairway discharge door at grade level. The *exit discharge* extends from the exterior door to the public sidewalk or public way.

The *UBC* specifically defines travel distance as the length of the exit path from any point within the building to an exterior door, **horizontal exit**, exit passageway door, or enclosed exit stairway door. The concept is the same as *exit access*.

Of the three elements (see Figure 7-1), only the exit access is regulated with

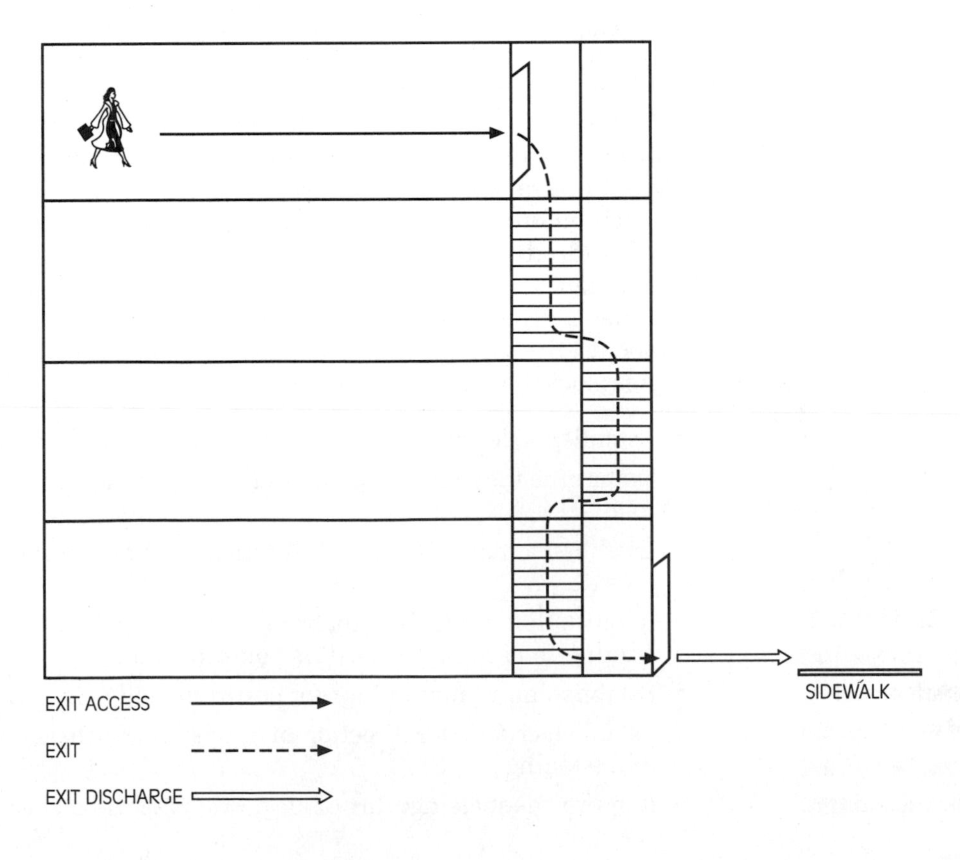

Figure 7-1 *The three parts of the means of egress.*

regard to travel distance. In essence, we are taking the viability of the exit for granted. We assume that fire prevention code provisions, which prohibit any storage in an exit and which require that all doors leading into the exit be maintained, self-closing, and positively latching, are diligently enforced. This clearly demonstrates the necessity of an effective fire prevention inspection program to ensure the reliability of a building's means of egress.

In small buildings, where the exterior door is within the maximum prescribed travel distance, exit access extends to the exterior door. The exit becomes the door itself and the exit discharge terminates at the sidewalk or public way.

design occupant load
the number of persons expected to occupy a space based on design tables in the building code (the number and capacity of exits is based on the design occupant load)

gross floor area
calculation used by the *BNBC* and *SBC* in determining the design occupant load of spaces with low occupancy densities, such as business, industrial, and mercantile areas; gross floor area includes the entire area within the exterior walls

net floor area
calculation used by the *BNBC* and *SBC* in determining the design occupant load of spaces with high occupancy densities, such as assembly and educational areas; net floor area includes the area within the walls exclusive of the thickness of walls or columns or accessory spaces such as stairs, rest rooms, or mechanical rooms.

floor area
calculation used by the *UBC* in determining design occupant load; floor area includes the area within the exterior walls exclusive of vent shafts, courts, and accessory spaces ordinarily used only by occupants of the main area

CODE PROVISIONS FOR ESTABLISHING MEANS OF EGRESS

Throughout this text we have identified the model building codes as design documents. Design professionals are able to work through the provisions in a logical order to incorporate the code provisions within their plan for a building. The same steps are used by the code official or fire inspector in assessing the adequacy of the means of egress from a building or space, or determining the occupant load in a given situation.

Step 1. Calculating the Design Occupant Load

The first step in determining the required means of egress capacity is to determine the number of persons who will occupy the space. Each of the model building codes has a similar table, which lists the use of the space and the square footage per occupant to be used (see Table 7-1).

It is important to note that the uses listed in the tables are not the *use group* of the building, but what the space is actually used for. The kitchen in a restaurant is not calculated as an assembly area, although the restaurant is a Group A building. It is an industrial area in the *BNBC* and *SBC* and has its own designation as a kitchen in the *UBC*.

In using the **design occupant load** (see Figure 7-2) concept, the code forces the designer to provide an adequate number of exits of the appropriate size for the number of people that should safely be able to use the space. The *BNBC* and the *SBC* use *net* and *gross floor area* when calculating design occupant load. **Gross floor area** refers to the square footage contained within the outside walls including hallways, closets, and rest rooms. **Net floor area** is the actual square footage that persons occupy. Deductions are made for stairways, columns and walls, rest rooms, mechanical rooms, and other accessory areas, leaving only the area to be physically occupied. *UBC* calculations are based on **floor area**, which is the space contained within the exterior walls, exclusive of vent shafts and courts. Accessory spaces such as rest rooms, ordinarily used only by persons occupying the main areas, are *not* included in the floor area calculation.[3]

Table 7-1 *Maximum floor area allowances per occupant.*

Occupancy	Floor area in Sq. Ft. per Occupant	Occupancy	Floor area in Sq. Ft. per Occupant
Agricultural Building	300 gross	Exercise rooms	50 gross
Aircraft hangars	500 gross	H-5 Fabrication and manufacturing areas	200 gross
Airport terminal		Industrial areas	100 gross
Concourse	100 gross	Institutional areas	
Waiting areas	15 gross	Inpatient treatment areas	240 gross
Baggage claim	20 gross	Outpatient areas	100 gross
Baggage handling	300 gross	Sleeping areas	120 gross
Assembly, gaming floors (keno, slots, etc.)	11 gross	Kitchens, commercial	200 gross
Assembly with fixed seats	See 1003.3.3.9 of the source document.	Library	
		Reading rooms	50 net
		Stack area	100 gross
Assembly without fixed seats		Locker rooms	50 gross
Concentrated (chairs only—not fixed)	7 net	Mercantile	
Standing space	5 net	Basement and grade floor areas	30 gross
Unconcentrated (tables and chairs)	15 net	Areas on other floors	60 gross
		Storage, stock, shipping areas	300 gross
Bowling Centers, allow 5 persons for each lane including 15 feet of runway, and for additional areas	7 net	Parking garages	200 gross
		Residental	200 gross
Business areas	100 gross	Skating rinks, swimming pools	
Courtrooms—other than fixed seating areas	40 net	Rink and pool	50 gross
		Decks	15 gross
Dormitories	50 gross	Stages and platforms	15 net
Educational		Accessory storage areas, mechanical equipment room	300 gross
Classroom area	20 net	Warehouses	500 gross
Shops and other vocational room areas	50 net		

Note: For SI, 1 square foot = 0.0929 m^2.

Source: 2000 International Building Code®, Table 1003.2.2.2, page 213. *Copyright 2000, International Code Council, Inc., Falls Church, Virginia. 2000 International Building Code®. Reprinted with permission of the author. All rights reserved.*

■ NOTE

In using the design occupant load concept, the code forces the designer to provide an adequate number of exits of the appropriate size for the number of people that should safely be able to use the space.

Once the design occupant load has been calculated, the designer can determine the number, location, and capacity of the exits from the space. All the model codes contain provisions to have the occupant load and seating plan posted or available on the premises for all assembly occupancies that have occupant loads exceeding fifty persons. Many occupancies contain multiple-use spaces such as hotel ballrooms, in which the occupant load for each use must be posted.

A typical ballroom may have two or more occupant loads posted. One occupant load is for unconcentrated assembly functions with tables and chairs, such

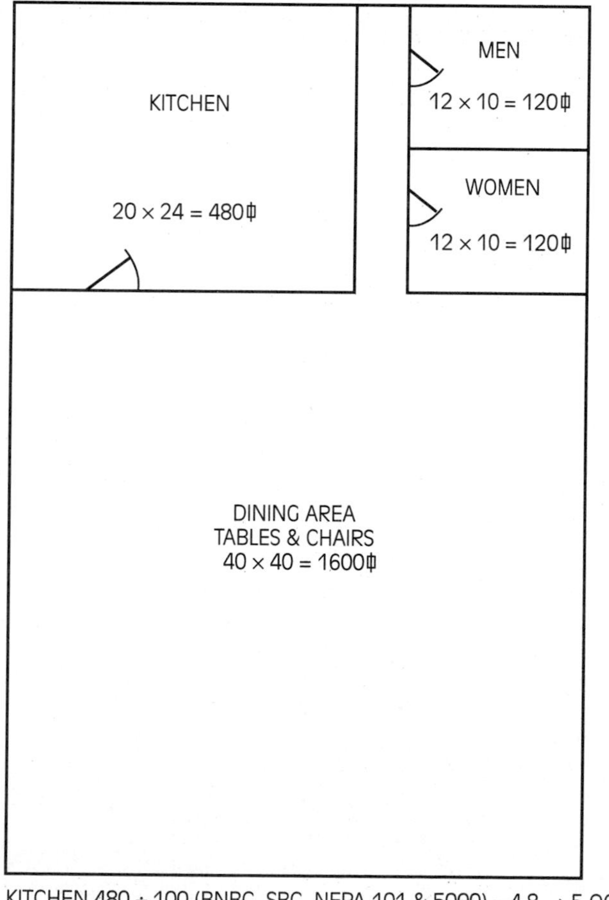

KITCHEN 480 ÷ 100 (BNBC, SBC, NFPA 101 & 5000) = 4.8 → 5 OCCUPANTS

(OR) 480 ÷ 200 (IBC & UBC) = 24 → 3 OCCUPANTS

DINING 1600 ÷ 15 = 106.6 → 107 OCCUPANTS

REST ROOMS ARE NOT INCLUDED BASED ON THE ASSUMPTION THAT THE OCCUPANTS HAVE BEEN INCLUDED IN THE KITCHEN OR DINING AREA CALCULATIONS

Figure 7-2
Determining design occupant load.

as banquets. The net floor area or floor area is divided by 15 square feet to determine the capacity. An occupant load is also established for concentrated assembly functions, with folding chairs, or no chairs at all. The floor area is divided by 7 square feet to determine the capacity of the ballroom (see Figure 7-3). The occupant load for the concentrated assembly use is more than double that of the unconcentrated use. In this example, the ballroom must have been constructed with exit facilities for the largest of the design occupant loads. Not only must the doors leading from the ballroom accommodate the larger number, but the corridors and all other components of the means of egress must also.

Step 2. Determining the Number of Exits Required

With a few exceptions, every building or floor area is required to be provided with at least two independent exits depending on the occupant load of the space. As shown in Table 7-2, all four model building codes have identical requirements for the number of exits based on occupant load.

Each of the model codes has a different method for determining when a building or floor area is only required to provide a single exit (see Table 7-3). Generally they are nonhazardous uses with low occupancy loads and relatively short travel distances.

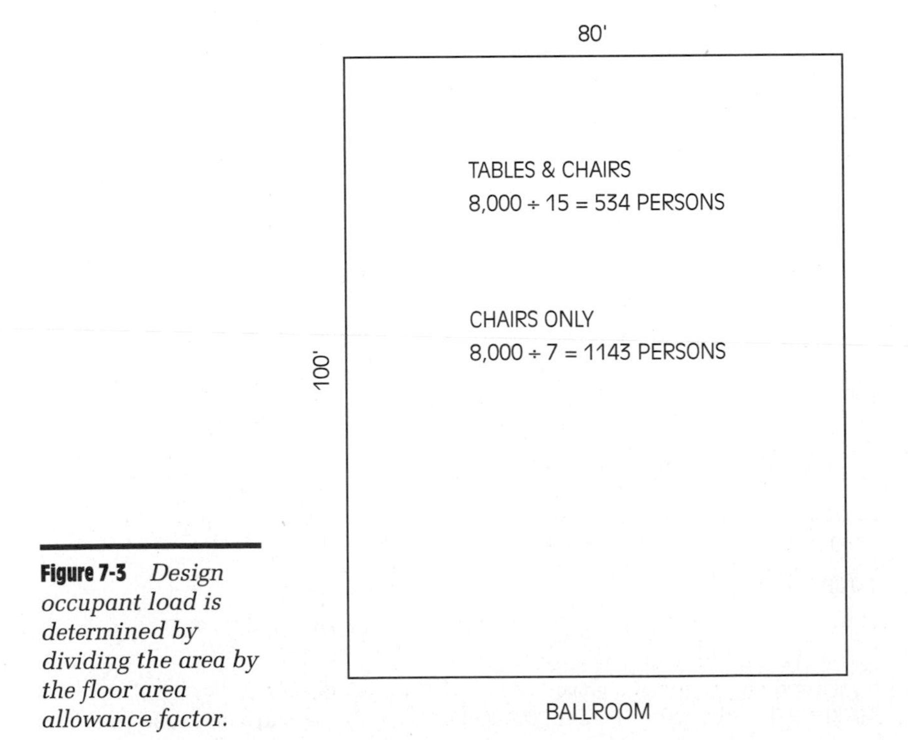

Figure 7-3 *Design occupant load is determined by dividing the area by the floor area allowance factor.*

Table 7-2 *Minimum required exits.*

Design Occupant Load	IBC	BNBC	SBC	UBC
500 or Fewer	2 Exits	2 Exits	2 Exits	2 Exits
501–1,000	3 Exits	3 Exits	3 Exits	3 Exits
Over 1,000	4 Exits	4 Exits	4 Exits	4 Exits
Code Section	1005.2.1	1010.2	1004.2.2	1003.1

Table 7-3 *Buildings with one exit.*

Occupancy	Maximum height of building above grade plane	Maximum occupants (or dwelling units) per Floor and Travel Distance
A, B[d], E, F, M, U	1 Story	50 occupants and 75 feet travel distance
H-2, H-3	1 Story	3 occupants and 25 feet travel distance
H-4, H-5, I, R	1 Story	10 occupants and 75 feet travel distance
S[a]	1 Story	30 Occupants and 100 feet travel distance
B[b], F, M, S[a]	2 Stories	30 occupants and 75 feet travel distance
R-2	2 Stories[c]	4 swelling units and 50 feet travel distance

Note: For SI, 1 foot = 304.8 mm.

[a] For the required number of exits for open parking structures, see Section 1005.2.1.1 [of the source document].

[b] For the required number of exits for air traffic control towers, see Section 412.1 [of the source document].

[c] Buildings classified as Group R-2 equipped throughout with an automatic sprinkler system in accordance with Section 903.3.1.1 or 903.3.1.2 and provided with emergency escape and rescue openings in accordance with Section 1009 shall have a maximum height of three stories above grade. [Section numbers refer to the source document.]

[d] Buildings equipped throughout with an automatic sprinkler system in accordance with Section 903.3.1.1 [of the source document] with an occupancy in Group B shall have a maximum travel distance of 100 feet.

Source: 2000 International Building Code®, Table 1005.2.2, page 236. *Copyright 2000, International Code Council, Inc., Falls Church, Virginia. 2000 International Building Code. Reprinted with permission of the author. All rights reserved.*

Step 3. Considering Exit Remoteness

Having enough exits to accommodate the number of occupants of a building or floor area is fine, but what happens when the fire or emergency situation causes one of the exits to become blocked? The number of persons trying to use the remaining exit or exits is doubled or worse; what if both exits are so close that the

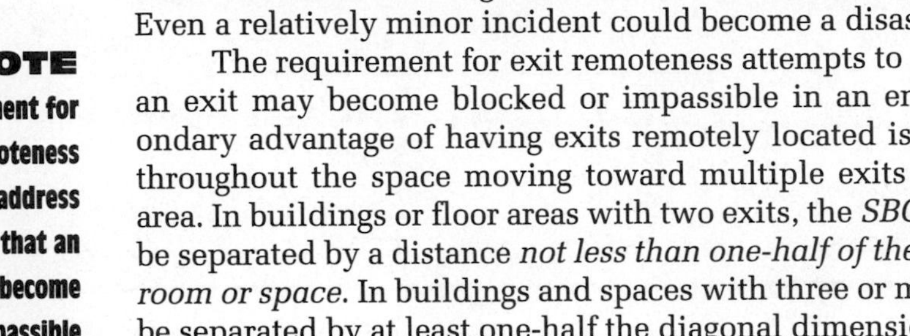

same problem affects them both? The fire that threatens the building next door becomes a whole new problem if the only exits from the exposed structure are on the same side of the building as the fire. Both exits could quickly become impassible. Even a relatively minor incident could become a disaster.

■ NOTE

The requirement for exit remoteness attempts to address the possibility that an exit may become blocked or impassible in an emergency situation.

The requirement for exit remoteness attempts to address the possibility that an exit may become blocked or impassible in an emergency situation. A secondary advantage of having exits remotely located is that occupants are spread throughout the space moving toward multiple exits and not congested in one area. In buildings or floor areas with two exits, the *SBC* and *UBC* require that they be separated by a distance *not less than one-half of the diagonal dimension of the room or space.* In buildings and spaces with three or more exits, at least two must be separated by at least one-half the diagonal dimension, and the others arranged to minimize the possibility of an incident blocking multiple exits.

The *BNBC* uses the same one-half the diagonal dimension requirement, but *reduces the distance in sprinklered buildings to one-fourth the diagonal dimension* (see Figure 7-4).

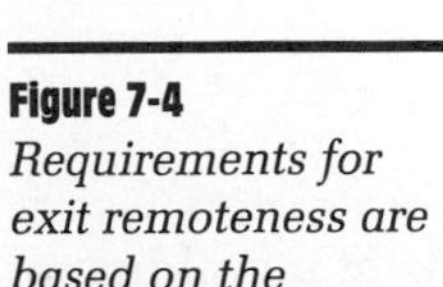

Figure 7-4
Requirements for exit remoteness are based on the diagonal dimension of the room or space.

Step 4. Limiting Maximum Travel Distance

■ NOTE
Limiting travel distance to an exit balances our desire to locate exits remotely with the need to enable persons to quickly reach an exit.

We have established the number of exits needed for a particular building or floor area based on design occupant load and ensured that they are adequately remote to guard against a single fire or emergency situation blocking multiple exits based on the size of the room or space. We are now faced with the possibility that the exits are so far apart, or the floor area so large, that the distance the occupants must travel before reaching an exit is too great. We have described our building regulations as a balancing act. Limiting travel distance balances our desire to locate exits remotely with the need to enable persons to quickly reach an exit.

The maximum distance that must be traversed before reaching a protected area such as an enclosed exit stair, exit passageway, exterior exit door, or horizontal exit or, in other words, the *length of exit access travel* is limited depending on the use group and whether the building is fully sprinklered.

dead end
a corridor, hallway, or passageway open to a corridor that can be entered from the exit access without passage through a door, but which does not lead to an exit

All four model building codes limit the length of **dead end** corridors or passageways to 20 feet. A dead end is defined by the *SBC* as *"a hallway, corridor or space open to a corridor so arranged that it can be entered from an exit access corridor without passage through a door, but does not lead to an exit."*[4] Twenty feet might not seem like much of a distance until you realize that for every step taken in a dead end, a return step must be taken after the occupants realize that they have gone the wrong way.

■ NOTE
Dead ends commonly occur in buildings undergoing renovation with partial occupancy.

Dead ends commonly occur in buildings undergoing renovation with partial occupancy. An exit might become unusable due to construction and another provided. The area of the building nearest the decommissioned exit has the potential for a dead end situation to be created. It is important to consider this element when assessing the viability of the means of egress in a building undergoing renovation with partial occupancy or in buildings where illegal, nonpermitted construction has taken place.

common path of travel
the portion of an exit access that building occupants must traverse before two distinct paths of travel to two exits are available

By regulating travel distance within the exit access, we are attempting to limit the amount of time that building occupants will be exposed to potential fire conditions. The *IBC,* NFPA 5000, and *SBC* take maximum travel distance a step further, by also limiting the travel distance before paths to two or more distinct exits become available. Regulating the **common path of travel** within the exit access limits the distance that building occupants must traverse before at least two distinct paths of travel to two or more exits become available.

Step 5. Calculating Egress Capacity

The only element lacking in establishing the exits from our building or floor area is how big we need to make them. How wide do the doors have to be? How about the corridors and stairs and the aisles that lead to them? All are addressed by the model building codes, depending upon the design occupant load and use group. The *BNBC* also makes allowances for fully sprinklered buildings. The design occupant load for the area served is multiplied by a factor to establish the width of the means of egress components.

The *UBC* simply states that the total occupant load served by the exit shall be multiplied by .3 for stairways and .2 for other exits. The *IBC, BNBC, SBC,* and NFPA 5000 provide tables with the appropriate factors. Use of these factors establishes a total exit capacity for the area served in inches, which is divided among the required number of exits, but not without other specific requirements to keep in mind.

Exit Stairways Serving Multiple Floors Where exit stairways serve multiple floors, the stairway capacity is determined using the floor with the highest occupant load, *not with the occupant load of the entire building.* The codes presume that in the time it takes for occupants to travel from the third floor landing to the second, the occupants of the second floor will have already made it to the first floor, and so on.

Where stairs converge, as in situations where exits from the basement and second floor meet and discharge on the first floor, the capacity of the exit from the first floor must be based on the occupant loads of the basement and the second floor *combined.*

Step 6. Ensuring Minimum Sizes for Means of Egress Components

For each element of the means of egress, there is a minimum size prescribed by the code, depending on the occupant load served (see Table 7-4). Regardless of the exit size based on capacity calculation, all components must conform with minimum size requirements. The net effect is that the minimum capacity calculated in inches is generally exceeded in order to conform with minimum size requirements. This is a plus for the safety of the occupants.

Table 7-4 *Minimum sizes for means of egress components.*

	Doors' Net Clear Opening	Corridors Serving ≤ 50	Corridors Serving > 50	Stairs Serving ≤ 50	Stairs Serving < 50
IBC	32	36	44	36	44
BNBC	32	36	44	36	44
SBC	32	36	44	36	44
UBC	32	36	44	36	44
NFPA 5000	32	36	44	36	44
NFPA 101					

Note: All sizes are in inches.

Corridors, Ramps, and Passageways

The required width of corridors ramps and passageways must be maintained. Projections such as drinking fountains or decorative trim are restricted depending on the model code, to a maximum of between 1¼ and 4 inches.

Doors that open into corridors are only permitted to reduce the corridor width by one-half and when in the fully open position, the door must be capable of folding against the corridor wall and projecting no more than 7 inches. Handrails are also limited to 3½ inches on each side of the corridor, or 7 inches total.

Fire-resistance Ratings for Corridors With some specific exceptions, corridors within occupancies that serve thirty people or more have to be rated for 1 hour. Those exceptions include fully sprinklered office buildings and a few others. The *BNBC* and *UBC* contain requirements for fire-resistance rated corridors within Chapter 10, Means of Egress. In the *SBC* the requirements are contained in Table 700 of Chapter 7, Fire-resistant Materials and Construction.

SMOKEPROOF AND PRESSURIZED ENCLOSURES

The travel distance to exit stairways within buildings is regulated, but once the occupants are within the exit enclosure they might have to descend three flights of stairs or thirty. The length of travel within the exit is *not* regulated. The model codes organizations recognized a need to ensure the viability of exits, particularly in high-rise buildings, where vertical shafts such as stairwells could act as chimneys for smoke and heat.

The exits from high-rise buildings (generally buildings with occupied floors more than 75 feet above the lowest level of fire department vehicle access) are given special attention by the codes. All exit stairs in high-rise buildings are required to be **smokeproof enclosures** or **pressurized enclosures**.

smokeproof or pressurized enclosure
an enclosed exit stair connected to all floors by either exterior balconies or ventilated vestibules and designed to limit the movement of smoke and fire gases into the stairwell

Smokeproof or pressurized enclosures are comprised of a 2-hour rated stairwell with rated opening protectives, connected to the floors of a high-rise by either exterior balconies that are open to the outside, or connected by ventilated vestibules. The stairshaft is also ventilated. The idea is to keep the smoke and heat out of the stairwell and vent it out the balcony or vestibule.

MAIN EXITS FROM ASSEMBLY BUILDINGS

Assembly buildings must have a main entrance and exit that discharges to the public way and is capable of accommodating one-half of the total occupant load. Additional exits must be provided to comply with requirements for exit remoteness, travel distance, exit capacity, and minimum exit size. The logic behind the requirement is that many of the occupants will immediately head for the main entrance in the event of an emergency. Although there are exits in every direction,

■ NOTE
Although there are exits in every direction, many people will insist on going out the way they came in.

many people will insist on going out the way they came in. They know that there is an exit there.

DOORS

With few exceptions, exit doors must be side swinging and open in the direction of egress when serving an occupant load of 50 or more or serving a high hazard occupancy. Panic hardware is required to release locks or latches on means of egress doors in assembly and educational occupancies, where occupant loads exceed 100, and in high hazard occupancies. Here we revisit the maintenance of rated assemblies discussed in Chapter 5. Panic hardware that is installed on fire doors must be labeled as *fire exit* hardware. Door hardware that is labeled fire exit has been tested under fire conditions. Hardware that is labeled *panic* has been tested strictly for panic and not fire. It has not been designed to be part of a fire resistance rated assembly. Look for the testing laboratories mark on the hardware, or look it up in the appropriate directory.

Revolving Doors

Revolving doors are an exception to the side swinging requirement. The death of 492 persons at the Coconut Grove fire in Boston in 1942 was due in large part to inadequate means of egress. Many exit doors were blocked or locked and the main exit, which most of the occupants were forced to attempt to use, was a revolving door.

The *UBC* permits the use of revolving doors, but they cannot be credited with providing any exit width when calculating exit capacity. The other codes limit credit for exit capacity to 50 percent or less. All five model building codes require that a conforming, side-swinging door be installed within 10 feet of every revolving door (see Figure 7-5). In order to be considered exits, they must also be capable of collapsing and forming parallel egress paths of at least 36 inches aggregate or total capacity.

■ NOTE
All five model building codes require that a conforming, side-swinging door be installed within 10 feet of every revolving door.

Horizontal sliding doors must be power operated with backup power supply, and be capable of being manually opened from both sides without special knowledge or effort in the event of power failure.

■ NOTE
Door width is determined by the actual clear opening provided and is not the physical measurement of the door.

Door Width

Door width (see Figure 7-6) is determined by the actual clear opening provided and is not the physical measurement of the door. A 3′0″ door (pronounced *three-oh*, for 3 feet 0 inches) is 36 inches wide when it comes from the factory. When it is installed in the jamb, the opening provided is reduced by the jamb itself, and sometimes by the hardware. That 3′0″ door probably provides a 34-inch clear opening. The minimum clear opening is 32 inches. Often, multiple doors or leaves

Figure 7-5 *A side swinging door must be installed within 10 feet of every revolving door. (Courtesy of Duane Perry.)*

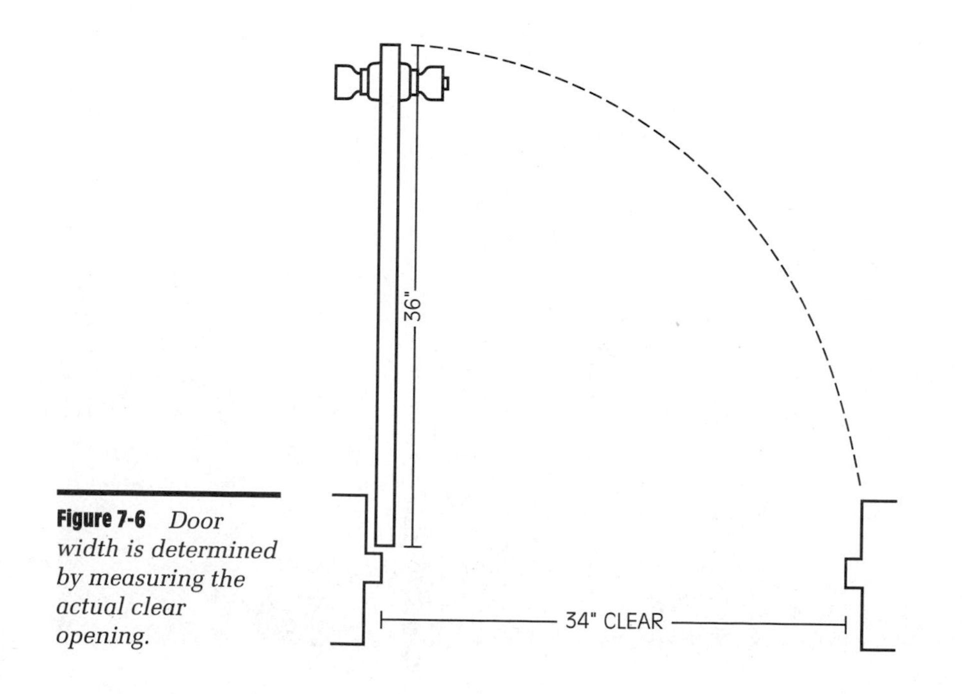

Figure 7-6 *Door width is determined by measuring the actual clear opening.*

are installed at an opening. The maximum width of a swinging door leaf is 48 inches. When a door opening has two leaves, as in Figure 7-7, at least one of them must provide a clear opening of 32 inches.

LOCKS AND LATCHES

The model building codes contain specific requirements for the locks and latches on egress doors. Unfortunately, these devices are often changed after tenant occupancy for security reasons, without regard to the original code requirements. The means of egress from a building or space may be totally adequate in location and capacity, but if the occupants can not get out the doors because of an unapproved locking or latching arrangement, the exits might as well not be there. Exit doors must be openable from the egress side without the use of a key or special knowledge or effort, with certain specific exceptions.

NOTE
Exit doors must be openable from the egress side without the use of a key or special knowledge or effort.

Business, Factory, Mercantile, and Storage occupancies are permitted by all the codes to install a key-operated lock on the egress side of the main egress door, provided a sign is posted stating that the door is to remain unlocked when the building is occupied.

Figure 7-7 *When door openings have two leaves, at least one must provide a 32-inch clear opening. (Courtesy of Duane Perry.)*

Special Locking Arrangements

The model building codes and the *Life Safety Code* (NFPA 101) have had significant differences in requirements for locking and latching arrangements on exit doors in areas with security concerns. NFPA 5000 includes means of egress requirements from NFPA 101. *A careful review of the requirements of your respective code is absolutely essential.*

■ NOTE
"Delayed egress locks" and "access-controlled egress doors" are two security arrangements addressed by the model building and fire codes and the *Life Safety Code*. They are drastically different systems with names that sound a lot alike.

"Delayed egress locks" and "access-controlled egress doors" are two security arrangements addressed by the model building and fire codes and the *Life Safety Code*. They are drastically different systems with names that sound a lot alike.

Delayed egress locks are time-delay, fail-safe locking systems. They are installed for security reasons, and have previously been permitted by the model building codes for all but Assembly and High Hazard uses, under certain conditions. The *BNBC* and *SBC* refer to delayed egress locks as "special locking arrangements," and only permit them in buildings with complete sprinkler or fire detection systems. The *UBC* refers to them as "special egress-control devices," and only permits them in buildings with *both* sprinklers and fire detection. The *2000 IBC* permits them in all but A, E, and H occupancies that are fully sprinklered or have fire detection systems. The *Life Safety Code* has permitted delayed egress locks in all but high hazard areas including assembly occupancies. All but the main entrance/exit doors may be equipped with delayed egress locks if the building is equipped with a sprinkler system or fire detection system.

Access-controlled egress doors are locked from both sides but are equipped with sensors on the inside that automatically unlock and permit persons to exit. This arrangement is permitted by the *IBC, BNBC, SBC, Life Safety Code,* and NFPA 5000. The *UBC* does not specifically permit the arrangement. The model building codes only permit the doors to be locked from the egress side after normal business hours in certain occupancies (A, B, E, and M in the *IBC*). Carefully check the provisions of your code!

EXIT ILLUMINATION AND EXIT LIGHTS

Except for dwelling units and inside of guest rooms in Residential occupancies, the means of egress in all buildings must be provided with lighting with a minimum intensity of 1 footcandle at the floor. This includes the exit access, exit, and *exit discharge*. Theaters and auditoriums are permitted to reduce that level to .2 footcandles during performances.

Illuminated exit signs and emergency lighting of the means of egress are required, depending on the code, in occupancies that exceed certain occupant loads or are required to have more than one exit. Emergency lighting is only required to provide 1 footcandle for a specified period (1–1½ hours), and is designed to enable the occupants to safely exit the building, *not to continue shopping or finish dinner.* All the codes contain an exit sign exception for buildings where the main entrance serves as an exit and is clearly visible. In these buildings, no exit sign is required at the main entrance/exit.

ACCESSIBLE MEANS OF EGRESS AND AREAS OF REFUGE

accessible means of egress
a means of egress including the exit access, exit, and exit discharge that can be entered and used by a person with a severe disability using a wheelchair and is also safe and usable for people with other disabilities

NOTE
Responsibility for the maintenance of *all* means of egress, accessible or nonaccessible, belongs to the fire official.

Each of the model building codes has provisions that require buildings to be accessible to persons with disabilities. Specific provisions for **accessible means of egress** are provided for in the codes. The installation of those elements from the building code are the responsibility of the building official. Responsibility for the maintenance of *all* means of egress, accessible or nonaccessible, belongs to the fire official.

The requirements are new to most fire officials and introduce some totally new concepts in occupant protection. *Unfamiliarity does not ease the burden of responsibility for enforcement of the maintenance provisions of the fire prevention code.* New buildings will be constructed with accessible means of egress complying with the model building codes. All the codes use the CABO/ANSI 117.1 standard *Accessible and Usable Buildings and Facilities* and each has an individual chapter with requirements for accessibility for people with disabilities.

With few exceptions, all occupied spaces must provide an accessible means of egress. If more than one exit is required, at least two accessible means of egress must be provided. Each must be continuous, and include corridors, ramps, exit stairs, elevators that comply with specific requirements, horizontal exits, and smoke barriers. In order to be considered a part of an accessible means of egress:

Exit stairs must have a clear width of 48 inches between handrails unless the building is fully sprinklered.

Elevators must be equipped with standby power. In nonsprinklered buildings, the elevator must be accessed through an area of refuge or horizontal exit.

Areas of refuge are spaces separated by 1-hour-rated smoke barriers. Doors must be rated for 20 minutes and equipped with approved hold-open devices that release upon smoke detector actuation. Each must be directly connected to an approved exit stairway or approved elevator and must have a two-way communications system.

ASSESSING THE EGRESS CAPACITY OF EXISTING SPACES

The same steps used in designing the means of egress for a building or space are used (in different order) to determine the exit capacity of an existing space. The concept of design occupant load, however, which we used to determine how many occupants our exits would have to accommodate, goes out the window. It is too late for that, the building is built. We are not interested in how many bodies can fit into a given area. Our concern is *how many bodies are the exits capable of accommodating.*

In a perfect world, fire inspectors would never encounter a building without the approved occupant load posted on the wall. The mayor would not hold his reelection kickoff reception for three hundred guests (including the fire chief) in his campaign chairman's newly constructed strip warehouse space. One thousand of the lovelorn would not show up for a book signing at the local 2,500 square-foot bookstore. You would never be placed in a position to determine "just how dangerous is this?"

Exit Capacity

The first step in determining the number of occupants a building or space can safely accommodate is to count the number of exit doors and determine the total net clear opening provided by them. The typical 3′0″ door is going to provide about a 34-inch clear opening. Estimate the clear opening widths of the exit doors, total the figures, and then divide by the appropriate factor from the building code. *The resulting figure is the total number of occupants that the exit doors can accommodate.*

The next step is to assess the means of egress components from the exit doors to the public way. Will the corridors accommodate the number of occupants that the doors did? Are there stairs? Measure the corridor and stair widths and divide by the appropriate factor in the building code. *The resulting figure is the total number of occupants that each of the components can handle* (see Figure 7-8).

Are the doors side swinging, and do they open in the direction of egress? Is there appropriate hardware on the doors for the use and number of occupants? How about exit remoteness and travel distance? In most cases, you will be able to determine a safe occupant load without much of a problem. Usually you will be dealing with a single room or space.

■ NOTE
Bars and unapproved security grills on emergency escape windows are a hazard to the public, and to firefighters.

■ NOTE
In order to install any system that affects egress from the structure, a building permit is required.

EMERGENCY ESCAPE WINDOWS

Most sleeping rooms in residential occupancies are required to provide at least one emergency escape window. The model codes vary a bit as to where they are required. The installation of security bars or other locking arrangements is prohibited unless the window can be completely opened without the use of tools or special knowledge or effort.

Bars and unapproved security grills on emergency escape windows are a hazard to the public, and to firefighters. In 1993 seven children died in a Detroit house fire and eight members of a family were killed in Mississippi due to barred windows. In 1995 twelve people died in three separate incidents in California and four people in a single fire in Brooklyn, New York.[5] In each case, the security devices prevented the occupants from escaping and proved to be an impediment to the fire department's rescue efforts.

The installation of such devices is regulated by the building codes and the fire prevention codes. The obstruction of a required means of egress is a clear vi-

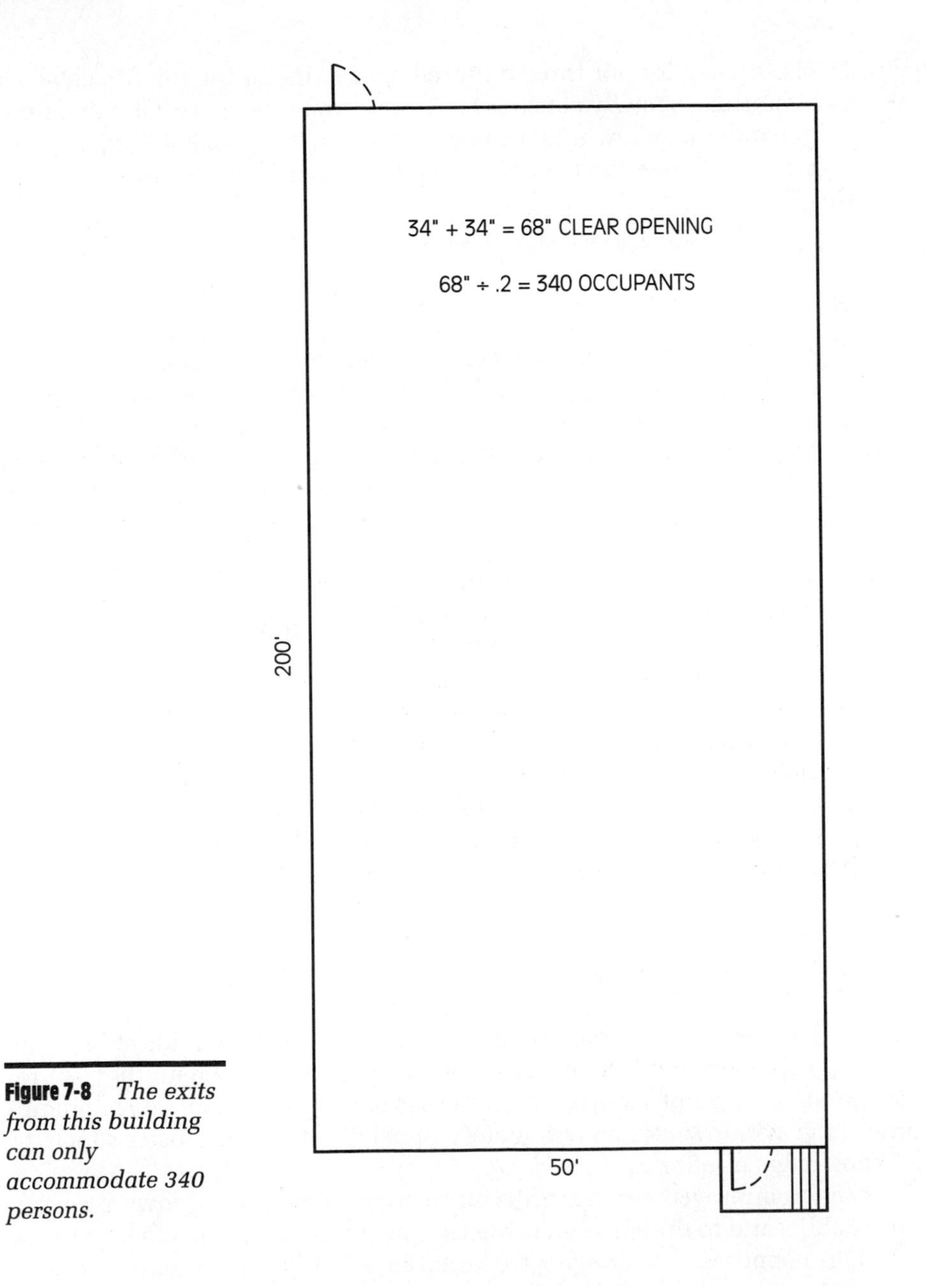

Figure 7-8 *The exits from this building can only accommodate 340 persons.*

olation of the fire prevention code. In order to install any system that affects egress from the structure, a building permit is required. If the system requires the use of tools or special knowledge or effort, the installation violates the building code and should not be approved.

■ NOTE
A fundamental principal of exiting is that building occupants can never be exposed to a higher degree of danger when passing from one element of the means of egress to another.

■ NOTE
Travel distance within the exit access is regulated to limit our exposure in the event of fire.

■ NOTE
The exit discharge is often overlooked by inspectors.

Summary

The model codes recognize three distinct elements of the means of egress—the *exit access*, the *exit*, and the *exit discharge*. A fundamental principal of exiting is that building occupants can never be exposed to a higher degree of danger when passing from one element of the means of egress to another. The most hazardous area is the exit access. Occupants are exposed to the building processes, combustible storage and furnishings, and ignition sources. Travel distance within the exit access is regulated to limit our exposure in the event of fire.

Travel distance within the exit is not regulated. Absolutely no storage of any type is permitted nor is the exit to be used for any purpose other than the movement of people. The exit discharge is often overlooked by inspectors. The area between the termination of the exit and the public way must be maintained free of snow, storage, or any other material that impedes safe egress. All portions of the means of egress must be illuminated at all times.

Components of the means of egress are designed based on the occupant load of the building or space. The design occupant load is based on the use of the building or space. Illegal changes in use can have a severe impact on the ability of the occupants to safely exit a structure if the building was designed to accommodate fewer occupants.

Review Questions

1. List the three elements of the means of egress.
   1. ________________
   2. ________________
   3. ________________
2. In which of the elements of the means of egress is travel distance regulated? ________
3. ________________ refers to the square footage contained within the outside walls including, hallways, closets, and rest rooms.
4. In calculating ________________, deductions are made for stairways, columns and walls, rest rooms, mechanical room, and other accessory areas.
5. Where exit stairs serve multiple floors, how is the stairway capacity determined? ________________
6. Where exit stairs converge from the second floor and basement and discharge on the first floor, how is the capacity of the first floor exit determined?

   ____
7. Main exits from assembly buildings must be

capable of accommodating ______________ of the total occupant load.

8. All the model building codes require that a ______________ be installed within 10 feet of every revolving door.

9. Door width is determined by measuring ______________.

10. In what occupancies do the model building codes require windows as a part of the means of egress? ______________

Discussion Questions

1. The owner of a local restaurant has installed an alarm lock device on his rear exit door. In order to exit, an occupant must first push or strike a red paddle clearly marked "push to open," then open the door using the approved panic bar.
   a. Does this configuration meet the requirements of your building code?
   b. What locking options are available for the owner?
2. All four model codes limit the distance the building occupants must traverse before entering a protected passageway or exiting the building. The *BNBC* and *SBC* define exit access within the code and then regulate travel distance within the exit access. The *UBC* more or less defines travel distance as the distance of the exit access.
   a. Why is the length of travel within the exit not regulated?
   b. Describe which building code provisions are designed to address this issue in covered mall buildings and high-rises.

Interior Finish Requirements

Learning Objectives

Upon completion of this chapter, you should be able to:

- Describe what building elements are regulated as interior finish.
- Describe the test method used to classify interior finish elements.
- Discuss the need to regulate interior finish based on previous fire experiences in the United States.

After the tragic 1942 Coconut Grove fire, in which 492 persons were killed and another 200 injured, a grand jury was convened. After 10 days of testimony by witnesses, the Suffolk County, Massachusetts, grand jury returned indictments against ten persons, including four officials from the city of Boston, the owner of the club, the wine steward, a contractor, a construction worker, and the interior decorator.[1]

Based on eye witness accounts that interior finish and decorative material played a significant role in the tragedy, Fire Commissioner William Reilly requested a chemist from the firm of Skinner and Sherman to test the material for flame resistance. The chemist's test report stated in part:

- The blue fabric from the molding in the Melody Lounge, where the fire reportedly started, burst into flame when touched with a match and was instantly consumed.
- Fibrous material wrapped around imitation palm trees burst violently into flames as a dry Christmas tree or excelsior would do.
- Straw and imitation straw matting did not ignite readily, but caught fire after a few seconds.
- The outer surface or coating of imitation leather wall covering from the lobby ignited readily and was consumed quickly. It was determined to be nitrocellulose.[2]

NOTE

Many incidents resulting in a large loss of life involved buildings with multiple deficiencies, where the lack of exits and highly combustible interior finish had a synergistic effect.

Table 8-1 lists fires in which interior finish played a significant role in a large loss of life and includes many names that are familiar to firefighters. Many of the incidents, such as the Coconut Grove, involved buildings with multiple deficiencies, where the lack of exits and highly combustible interior finish had a synergistic effect.

Flame-resistance specifications for interior finish and decorative materials are spelled out in the model building codes. Provisions intended to maintain that level and to regulate decorative materials brought into the structure are included in the fire prevention codes.

Table 8-1 *Fires where interior finish played a significant role.*

Fire	Location	Date	Fatalities
Coconut Grove	Boston, MA	11/28/42	492
LaSalle Hotel	Chicago, IL	6/05/46	61
Our Lady of Angels School	Chicago, IL	12/01/58	95
Beverly Hills Supper Club	Southgate, KY	5/28/77	164
MGM Grand Hotel	Las Vegas, NV	11/21/80	85

INTERIOR FINISH AND TRIM

Interior finish and trim means the exposed interior surfaces of a building, meaning the ceiling, walls and partitions, columns, floor finishes, wainscoting, paneling or any other finish applied structurally for insulation, fire resistance, acoustical, or decorative purposes. Interior finish contributes to fire development and intensity in four ways:

1. It affects the rate at which flashover occurs.
2. It contributes to fire extension.
3. It may intensify the fire by contributing fuel.
4. It may produce smoke and toxic gases.[3]

Interior finish materials other than floor finish must be classified in accordance with ASTM E84, *Test Method for Surface Burning Characteristics of Building Materials*; NFPA 255; UL 723; or UBC Standard 8-1. All are designations of the **Steiner tunnel test**, developed by A. J. Steiner, an engineer at Underwriters' Laboratories.

Steiner tunnel test test to determine the surface burning characteristics of building materials in which the flame spread of the test material is compared to asbestos cement board, rated 0, and red oak, rated 100; the higher the rating, the greater the potential hazard.

In the Steiner tunnel test, materials are exposed to gas jets in a 25-foot horizontal furnace. Flame spread in the test material is then compared to asbestos cement board, which has an assigned flame spread rating of 0, and red oak, which has an assigned rating of 100. The higher the rating, the greater the potential hazard. Three physical characteristics are measured during the test: flame spread, smoke development or smoke density, and fuel contributed to the fire. As shown in Table 8-2, the *BNBC* and *UBC* use the numerals I through III, and the *IBC, SBC,* and NFPA 5000 and 101 use the letters A through C to designate materials.

As indicated in Table 8-3, the model building codes regulate flame spread by specifying ratings according to use group classification and location within the means of egress. In all cases, smoke development must be no greater than 450 for all materials.

Table 8-2 *Flame spread ratings.*

Flame Spread	*IBC*	*BNBC*	*SBC*	*UBC*	NFPA 5000 & 101
0–25	A	I	A	I	A
26–75	B	II	B	II	B
76–200	C	III	C	III	C

Table 8-3 *Interior Wall and Ceiling Finish Requirements by Occupancy*[k].

GROUP	SPRINKLERED[l]			UNSPRINKLERED		
	Vertical exits and exit passageways[a,b]	Exit access corridors and other exitways	Rooms and enclosed spaces[c]	Vertical exits and exit passageways[a,b]	Exit access corridors and other exitways	Rooms and enclosed spaces[c]
A-1, A-2	B	B	C	A	A[d]	B[e]
A-3[f], A-4, A-5	B	B	C	A	A[d]	C
B, E, M, R-1, R-4	B	C	C	A	B	C
F	C	C	C	B	C	C
H	B	B	C[g]	A	A	B
I-1	B	C	C	A	B	B
I-2	B	B	B[h,i]	A	A	B
I-3	A	A[j]	C	A	A	B
I-4	B	B	B[h,i]	A	A	B
R-2	C	C	C	B	B	C
R-3	C	C	C	C	C	C
S	C	C	C	B	B	C
U		No restrictions			No restrictions	

Notes: For SI, 1 inch = 25.4 mm, 1 square foot - 0.0929 m^2.

[a] Class C interior finish materials shall be permitted for wainscotting or paneling of not more than 1,000 square feet of applied surface area in the grade lobby where applied directly to a noncombustible base or over furring strips applied to a noncombustible base and fireblocked as required by Section 803.3.1 [of the source document].

[b] In vertical exits of buildings less than three stories in height of other than Group I-3, Class B interior finish for unsprinklered buildings and Class C interior finish for sprinklered buildings shall be permitted.

[c] Requirements for rooms and enclosed spaces shall be based on spaces enclosed by partitions. Where a fire-resistance rating is required for structural elements, the enclosing partitions shall extend from the floor to the ceiling. Partitions that do not comply with this shall be considered enclosing spaces and the rooms or spaces on both sides shall be considered one. In determining the applicable requirements for rooms and enclosed spaces, the specific occupancy thereof shall be the governing factor regardless of the group classification of the building or structure.

[d] Lobby areas in A-1, A-2, and A-3 occupancies shall not be less than Class B materials.

[e] Class C interior finish materials shall be permitted in places of assembly with an occupant load of 300 persons or less.

[f] For churches and places of worship, wood used for ornamental purposes, trusses, paneling, or chancel furnishing shall be permitted.

[g] Class B material required where building exceeds two stories.

[h] Class C interior finish materials shall be permitted in administrative spaces.

[i] Class C interior finish materials shall be permitted in rooms with a capacity of four persons or less.

[j] Class B materials shall be permitted as wainscotting extending not more than 48 inches above the finished floor in exit access corridors.

[k] Finish materials as provided for in other sections of this code.

[l] Applies when the vertical exits, exit passageways, exit access corridors or exitways, or rooms and spaces are protected by a sprinkler system installed in accordance with Section 903.3.1.1 or Section 903.3.1.2 [of the source document].

Source: 2000 International Building Code®, Table 803.4, page 173. *Copyright 2000, International Code Council, Inc., Falls Church, Virginia. 2000 International Building Code. Reprinted with permission of the author. All rights reserved.*

APPLICATION

Interior finish applied to walls and ceilings must be applied directly to the structural elements or attached to furring strips. If furring strips are used, the void spaces created must be filled with inorganic or Class I materials, or fireblocked.

CARPET AND CARPETLIKE MATERIALS

Carpet and carpetlike materials that are installed as wall coverings pose a particular hazard. Less than 3 months after the disastrous MGM Grand Hotel fire, another Las Vegas hotel, the Las Vegas Hilton was struck by a fire that involved twenty-two floors of the thirty-story building. The fire was set with a small open flame device by a motel employee in an elevator lobby on the eighth floor.

The fire quickly involved the furnishings and the *carpeting, which covered the walls and ceiling.* After flashover occurred, the elevator lobby window failed, allowing a flame front to extend upward on the exterior of the building. Radiant heat exposure led to involvement and extension on upper floors, due in large part to the carpeted walls and ceilings of the elevator lobbies. Firefighters on the scene estimated that vertical extension to the top of the building took only 20–25 minutes.[4]

NOTE
Carpeting manufactured as a floor covering is tested in a horizontal position, as it would normally be installed; the results of that test may have no relation to the actual burning characteristics of the carpet installed vertically on a wall.

Carpeting that is manufactured as a floor covering is tested in a horizontal position, as it would normally be installed. The results of that test may have no relation to the actual burning characteristics of the carpet when installed on a wall vertically. All three model building codes require that carpet and carpetlike wall coverings shall either:

- Be Class I or Class A (0–25) and be installed in rooms or areas protected with automatic sprinklers *or*
- Meet the acceptance criteria of a full-scale room fire test specified by the individual codes.

Although all three model codes have contained the requirement for some time, many carpet suppliers may not be aware of the requirement.

DECORATIVE MATERIALS

Decorative materials are treated differently from code group to code group. All of the fire prevention codes have specific requirements that we address in Chapter 9.

The *IBC* and *BNBC* have specific requirements for decorative materials: All decorative material in assembly and institutional occupancies and hotel/motels must be noncombustible or flame resistant. They do not limit the amount of noncombustible decorative material that can be installed, but limit treated flame-resistant material to 10 percent of the total wall and ceiling area.

Summary

The interior finish and trim and materials used for decorative purposes can pose significant hazards to building occupants. The model code regulations for these items were developed because of the severe fire potential of certain materials. Fires in which these materials contributed to rapid fire development and rapid fire spread have often resulted in numerous fatalities and injuries.

Review Questions

1. What is meant by *interior finish*?

 ____

2. List four ways in which interior finish contributes to fire development and intensity.

 1. ______________________
 2. ______________________
 3. ______________________
 4. ______________________

3. ASTM E84, UL 723, and UBC Standard 8-1 are all designations of the ______________________ Test.

4. The test apparatus used in the ASTM E84, UL 723, and UBC Standard 8-1 is a furnace that tests materials in a ______________________ position.

5. List the three characteristics that are measured during the ASTM E84, UL 723, and UBC Standard 8-1 Tests.

 1. ______________________
 2. ______________________
 3. ______________________

6. According to your model building code, what is the classification of a material that has a flame-spread rating of 10? ______________________

7. In a sprinklered building, what is the minimum flame-spread rating for carpeting that is to be installed as a wall covering? ______________________

8. What is the maximum smoke development permitted for interior finish materials?

 ____

9. Name two major fires in which interior finish was a significant factor.

 1. ______________________
 2. ______________________

10. Is exposed insulation within an occupied space considered interior finish? ______________________

Discussion Question

1. You receive a call from the chief. He has been approached by the local church whose request to install carpet on the walls of their unsprinklered sanctuary was denied by the building official, based on your recommendation. The chief voices the same concern as the church board of directors: "It is Class I, the safest possible rating."

   a. Explain the basis for your decision to the chief.

   b. Would the installation of smoke detectors in the sanctuary be a satisfactory solution?

Section 3

THE FIRE PREVENTION CODE

In the introductions to Sections 1 and 2, we discussed the importance of the roles and responsibilities of code officials and enforcing agencies. Hybrids, in which multiple disciplines are the responsibility of a single agency, have become more common as governments reorganize and consolidate. More and more code officials wear several hats and enforce several different codes, a situation that forces all of us to constantly bear in mind our legal responsibilities, duties, and authority.

For the remainder of the text, get out your fire official hat. We will discuss the *safeguarding of life and property from fire and explosion arising from the storage, handling, and use of hazardous substances, materials, and devices, and from conditions hazardous to life and property in the use or occupancy of buildings or premises,*[1] through the use of the technical provisions of the fire code. You will see very quickly that the fire prevention code does a lot more than just keep fires from occurring. It contains provisions for minimizing the impact of the fires that occur despite our best efforts at prevention.

General Fire Safety Provisions

Learning Objectives

Upon completion of this chapter, you should be able to:

- Describe NFPA 550, *The Fire Safety Concepts Tree*, and its uses.
- List the two strategies for fire safety used by the model fire prevention codes.
- List three objectives for achieving the goal of life and property protection using each strategy.
- Describe the significance of illegal changes in use and list the code sections that apply.
- List four rated assemblies that an inspector would routinely check to ensure compliance with the code.

GENERAL PROVISIONS FOR ALL OCCUPANCIES

Successful fire prevention is a bit like football. You can have all the razzle-dazzle plays in the world, but if your team cannot block on the line and cannot tackle, you will not win. These are not recent revelations. Lombardi said it, Landry said it, and every successful NFL coach is still saying it. Forget the basic fundamentals, and no matter how complex your system, you will not overcome your basic weaknesses.

■ NOTE

The fundamentals of fire protection and prevention have not changed since A.D. 64, when Emperor Nero was said to have fiddled while Rome burned.

Despite all the technological breakthroughs that are occurring at an accelerating rate, the fundamentals of fire prevention and protection have not changed since A.D. 64, when Emperor Nero was said to have fiddled while Rome burned. We have not come up with anything yet to replace what we have learned the hard way over and over since the very beginning of our civilization: (1) Eliminate ignition sources, (2) control the available fuel, and (3) prepare for the times you fail at numbers 1 and 2.

NFPA 550, *Guide to Fire Safety Concepts Tree* (see Figure 9-1), was first developed in 1974 and identifies fundamental strategies and lists objectives for achieving a fire safe environment. The fault tree analysis system was first developed by H. A. Watson of Bell Telephone labs in 1962 and made famous by Boeing Corporation in its application to the Minuteman ballistic missile program.[1]

The *Fire Safety Concepts Tree* is the basis for the U.S. Fire Administration's interactive course/reference Fire Safe Building Design[2] on CD-ROM. The program is de-

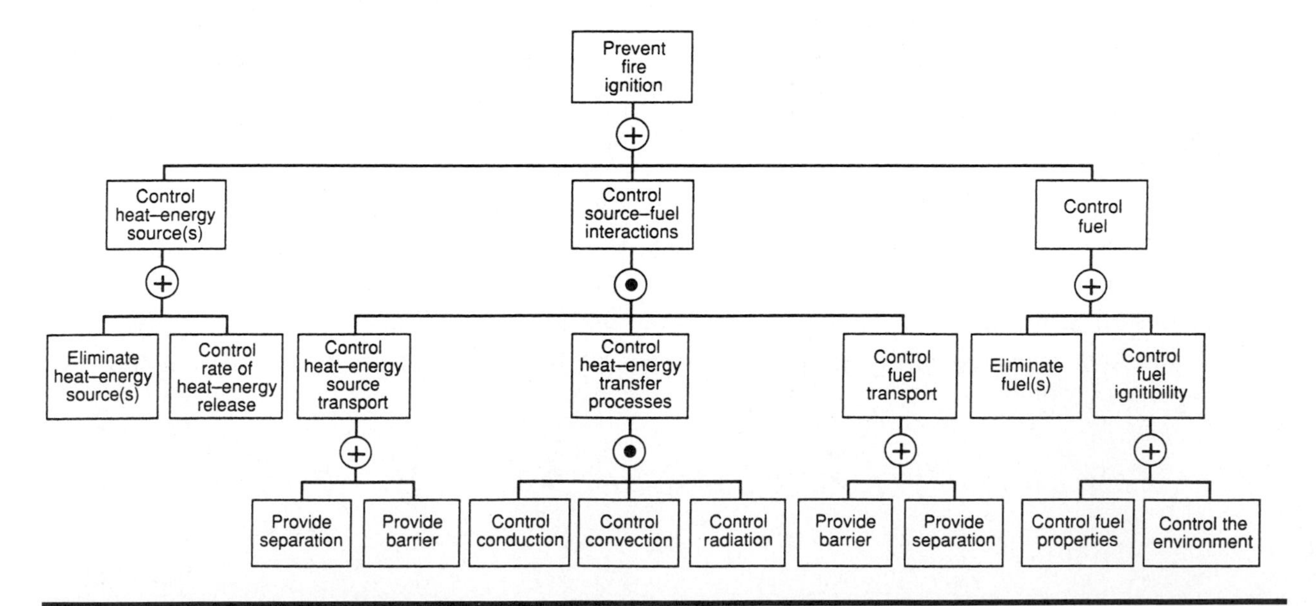

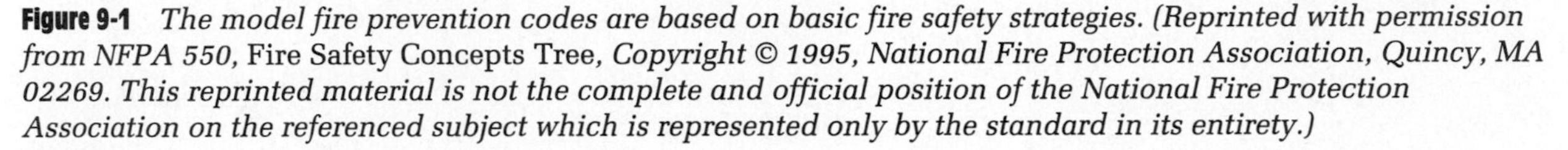

Figure 9-1 *The model fire prevention codes are based on basic fire safety strategies. (Reprinted with permission from NFPA 550,* Fire Safety Concepts Tree, *Copyright © 1995, National Fire Protection Association, Quincy, MA 02269. This reprinted material is not the complete and official position of the National Fire Protection Association on the referenced subject which is represented only by the standard in its entirety.)*

signed for architects and designers and does an excellent job of using the decision tree to explain the fundamentals of fire safe design and construction, as well as the fundamentals of fire prevention. The program also provides valuable insight for the inspector whose job is the protection of those design features for the life of the building.

The goal of the model fire and building codes, which is the safeguarding of life and property, is achieved through two basic strategies: *to prevent fires from occurring*, and, when they do occur through some failure, *to manage the impact* on people and property.

Consider the building code fundamentals discussed in Chapters 3 through 8. Regulating use and occupancy, requiring fire-resistive construction elements and built-in fire protection systems, regulating interior finish, and requiring adequate egress facilities are all functions of the model building codes. These features prescribed by the model building codes do not keep fires from starting. *They are aimed at minimizing the impact of fire.*

Fire prevention codes contain provisions to keep hostile fires from occurring, as well as those that attempt to minimize the impact of fires by requiring the maintenance of the fire safety construction features required by the building codes.

Prevent Ignition

NOTE
Prevent the ignition of fires by effectively segregating the three elements of the fire triangle.

Prevent the ignition of fires by effectively segregating the three elements of the fire triangle.

- Control *heat* sources such as smoking, unsafe heating appliances, electrical hazards, open flames, and burning.
- Control *fuels* by regulating the storage and handling of specific materials.
- Control the *interaction* of fuel, oxygen, and the heat source in storage, handling, transportation, and processing.

Manage Impact

NOTE
Manage the fire impact by effectively regulating construction, use and occupancy, and occupant training.

Manage the fire impact by effectively regulating construction, use and occupancy, and occupant training.

- *Limit* building height and area based on the combustibility and fire resistance of construction elements.
- Provide adequate *fire resistance* for structural elements and separation assemblies.
- Provide adequate *emergency egress* facilities.
- Provide built-in *fire protection systems* to detect and suppress fires.
- Ensure that hazardous processes and occupancies are adequately *separated* from other occupancies.
- Ensure that occupants and employees in critical occupancies receive appropriate *training* in evacuation and emergency response.

As we study the provisions of the fire prevention codes, trace each one back to the basic fire prevention fundamentals. These same fundamentals were behind the development of most nonmodel code-related fire prevention ordinances adopted by state and local governments over the years.

CHANGES IN USE AND OCCUPANCY

Of all the fire prevention code violations that you will ever encounter, the *illegal change of use* presents the largest potential for the number of individual safety deficiencies that can exist. As we learned in Section 2, each use group has specific requirements that range from limits on building height and area and the installation of fixed fire protection to the establishment of means of egress. An illegal change in use can take a building that conforms to the requirements of the original use group and create a situation in which the means of egress and fire protection features are totally inadequate, often to the surprise of the occupants.

The 15,000-square-foot Business Use Group building that is illegally converted into a house of worship, easily results in an occupant load that is more than 14 times greater than that which was used to design the exits for the original Use Group B building! The original use did not require a sprinkler system, fire alarm system, or panic hardware on exit doors. Now more than 2,000 people have congregated in a building with only enough exits to quickly evacuate about 150 people, and without the fire protection features that years of experience have proved are vital to prevent catastrophes. What is the violation? Should the inspector cite a violation of Chapter 10, Means of Egress, of the building code? How about Chapter 9, Fire Protection Systems?

The violation that the *fire code official* can cite in this instance is "illegal change of use," prohibited by the code provisions shown in Table 9-1 by all five model fire prevention codes.

The lack of adequate means of egress or inadequate fixed fire protection may be the overriding safety concerns, but the actual violation of the fire prevention code is the illegal change in use. The fire official's power is limited to (1) issuing a Notice of Violation or order prohibiting the illegal use or activity, (2) taking immedi-

Table 9-1 *Illegal changes in use violation.*

Code	Section
2000 International Fire Code	102.3
2000 NFPA 1 Fire Prevention Code	1-9.4.1
1999 BOCA/National Fire Prevention Code	301.2
1999 Standard Fire Prevention Code	102.5.2
2000 Uniform Fire Code	103.3.2.3

ate steps to ensure the safety of the occupants, including issuing an emergency evacuation order if warranted, and (3) reporting the condition to the building official.

Issuing a violation under the building code will probably get your notice overturned in court or before the appeals board, unless you are with one of the fire prevention bureaus specifically empowered to enforce the building code. Remember what hat you are wearing!

■ NOTE

Hazardous materials include physical and health hazards regulated by the codes, and run the gamut from combustible liquids to baled wastepaper and ammonium nitrate fertilizer.

Other illegal changes in use can be more subtle, such as from low to moderate hazard storage, or from one assembly use to another. The quantities of regulated hazardous materials stored or handled within a building can also create an illegal change in use. The term *hazardous materials*, as used within the model code process, is not limited to the chemicals that only the members of the hazmat team can spell. Hazardous materials include physical and health hazards regulated by the codes, and run the gamut from combustible liquids to baled wastepaper and ammonium nitrate fertilizer. They are the same items that fire prevention codes regulated before the terms *hazardous materials* and *hazmat* became popular. The storage or use of these substances in quantities that exceed a prescribed threshold, or *exempt quantity*, constitute an illegal change in use. There is more on hazardous materials code requirements in Chapter 12.

Perhaps one of the greatest challenges facing fire and building officials as well as fire departments is the growing number of senior citizens who need to live in a structured environment, yet fall somewhere short of being "incapable of self-preservation." Structures that do not have the built-in fire protection features required for hospitals and nursing homes are being used more and more to house elderly persons, who may or may not be capable of appropriate response and evacuation in the event of a fire emergency. Political expediency, political correctness, and short-sighted economics all tend to point toward lax enforcement of the fire prevention code in this instance.

Unwillingness on the part of some fire officials to rectify illegal changes in use when facilities are used for nonambulatory residents and do not meet the strict requirements for institutional occupancies will inevitably lead to fires involving injuries and deaths.

The use group classification system of the BOCA National Building Code was the subject of a lawsuit brought under the Americans with Disabilities Act and Federal Fair Housing Act within the federal court system in 1995.[3] At issue was the use group classification system's use of disability (capacity for self-preservation) as a means to determine occupancy classification. The case was dismissed, but others will inevitably follow. Is it better to be sued for a good faith effort at enforcing a legally adopted, nationally recognized model code; or is it better to be charged for criminal nonfeasance and be sued after a disastrous fire for *not* enforcing it?

MAINTENANCE OF RATED ASSEMBLIES AND BARRIERS TO FIRE

The model building codes take a few things for granted. Some of them, like gravity, are sure bets. Buildings try to fall down, not up. Others, like the continued

Figure 9-2 *This rated exit enclosure has been compromised during construction.*

■ **NOTE**

Simply put, all required fire-resistance-rated assemblies and required barriers to fire must be maintained at all times.

viability of tenant separation walls, rated corridors (see Figure 9-2), and exit enclosures are not such good bets, unless there is an ongoing, aggressive inspection program that ensures their survival. Each of the model fire prevention codes adequately covers the subject in less than a page. Simply put, all required fire-resistance rated assemblies and required barriers to fire must be maintained at all times. This list includes walls, partitions, floor/ceiling assemblies, stairway enclosures, draftstopping, fireblocking, firestopping, column protection, and *all other rated assemblies* and all opening protectives.

■ **NOTE**
A primary consideration during every inspection must be whether required ratings have been violated by illegal construction or the addition of improper equipment or hardware.

A primary consideration during every inspection must be whether required ratings have been violated by illegal construction or the addition of improper equipment or hardware. This determination is not possible if the inspector does not know whether rated assemblies were required in the first place, so the building code must always be within arms' reach.

Rated Walls, Partitions, and Floor/Ceiling Assemblies

approved
acceptable to the code official with jurisdiction; approval is normally based on nationally recognized standards or, in their absence, on sound engineering practice

Rated walls, partitions, and floor/ceiling assemblies can be illegally penetrated in order to combine two tenant spaces, run utilities such as communications cables, or for a host of other reasons. The code is explicit that repairs must restore the assembly to its original rating through an **approved** method. A cosmetic sheetrock repair, in which a partition is merely patched and may look fixed, is not sufficient.

■ **NOTE**
Repairs must restore the assembly to its original rating through an approved method; a cosmetic sheetrock repair, in which a partition is merely patched and may look fixed, is not sufficient.

Inspections of corridor walls and tenant separation walls often requires a look above the ceiling to determine that walls extend to the roof deck. If the area above the ceiling is used as a return air plenum and the building is of noncombustible construction, this is the time to check that any cable installed is listed or labeled as being approved for plenum use.

Structural Elements

Ratings for structural elements such as columns and beams are particularly prone to damage in warehouses and buildings where occupants have access and are forced to work around the assemblies. Employees operating forklifts or other machinery can damage protective assemblies of gypsum or spray-applied fire-resistant material. You can order the repair of the assembly at every inspection, and come back a week later to find that it has been damaged again. Impact protection, with metal corner guards or a substantial jacket of metal or other noncombustible material should be required and considered an integral part of the rated assembly.

■ **NOTE**
Impact protection, with metal corner guards or a substantial jacket of metal or other noncumbustible material should be required and considered an integral part of the rated assembly.

Wires, cables, and pipes are not permitted to be embedded in the fire protective covering of a structural member that is individually encased. Likewise, gypsum or spray-applied fire-resistant material cannot be penetrated by tradesmen attempting to use the protected structural element for some purpose.

Draftstopping and Fireblocking

Draftstopping and fireblocking are barriers to limit fire spread that are *not rated assemblies*. The building code requirements for these two components are prescriptive and are not based on performance. Most fireblocking will be hidden from view behind wall and ceiling construction. Draftstopping, especially in apartment

■ NOTE
Although draftstops are not rated assemblies, the model building codes require that their "integrity be maintained."

buildings of combustible construction, is a critical element that should be checked during the inspection. Although draftstops are not rated assemblies, the model building codes require that their "integrity be maintained."

Builders have the option of providing an approved access opening to each draftstopped area or providing openings within the draftstopping. Openings in the draftstops must be self-closing and positively latching. Access openings in rated ceilings must also be protected with listed opening protectives.

Opening Protectives

■ NOTE
The label is a required component of the assembly, just as the latching mechanism of a fire door is a required component.

Opening protectives such as fire doors, fire windows, and fire dampers are required to be accessible for inspection and labeled at the factory with a label or listing mark indicating the fire protection rating. If the label or mark is not there, it is not a rated assembly. The label is a required component of the assembly, just as the latching mechanism of a fire door is a required component. If there is no label on the opening protective, it must be replaced.

Side-swinging fire doors and smoke doors must be self-closing and positively latching. Doors must close and latch when released from any position, from fully opened to nearly closed. The installation of hardware such as special locks, mail slots, and door alarms are prohibited unless the hardware is listed for installation in the door. NFPA 80, *Fire Doors and Windows*, contains specific provisions regarding the modification of fire doors.

■ NOTE
Door stops, wedges, and other hold-open devices are prohibited on fire doors.

Door stops, wedges, and other hold-open devices are prohibited on fire doors (see Figure 9-3). Approved hold-open devices that are self-closing upon the activation of automatic fire detectors are permissible, but must be installed under a building permit.

Rolling fire doors (see Figure 9-4) must be tested annually by the building owner and records maintained on the premises. If you do not ask for the records during your inspection, you can count on two things: (1) they are not keeping the records and (2) they are also not testing the doors. If you wait until the fire occurs in which something awful happens and then ask for the records, you are too late. Fusible links that operate rolling fire doors must be kept clean and unpainted.

■ NOTE
Although interior finish materials are seldom the first material ignited in a building fire, wall and ceiling finish plays a major role in fire development and fire spread.

DECORATIVE MATERIAL AND INTERIOR FINISH

In Chapter 8 we discussed the role that combustible interior finish materials have played in major fires and how the model building codes were changed to address the problem. Although it is closely regulated by the model building codes, interior finish continues to play a significant role in fatal fires.

Decorative wall coverings are often installed without the benefit of the building permit process, either through ignorance or by unscrupulous contractors who choose to ignore the regulations. Carpetlike textile wall coverings, wood paneling, and even plastic decorative panels are installed over noncombustible gypsum wallboard. This significantly adds to the fuel load, increases potential fire spread,

Figure 9-3 *Illegal hold-open devices come in many shapes and sizes. (Courtesy of Ron Berry.)*

and accelerates flashover. Although interior finish materials are seldom the first material ignited in a building fire, wall and ceiling finish plays a major role in fire development and fire spread.[4]

After a 1990 fraternity house fire in Chapel Hill, North Carolina, NFPA's senior fire investigator identified the building's 1 inch by 8 inch floor-to-ceiling, southern pine paneling as a major culprit in the five deaths that occurred.[5] The paneling had been installed on furring strips that had been nailed to the building's plaster walls. "If it hadn't been for the wood paneling, which created a very intense fire in the basement, this tragedy might not have happened."[6] The *IFC* requirements for interior finish in existing buildings are shown in Table 9-2.

Decorative Materials

Decorative materials that may change with the seasons are also closely regulated by the model fire prevention codes. The *UFC* even has specific procedures for Christmas trees in public buildings. *UFC* Appendix IV-B regulates the flame retardance, support device, watering, and use of electric lights. The regulations are a useful guide no matter which model code you use. The potential fire problem associated with Christmas trees cannot be overstated. In 1990 alone, there were 29 deaths, 100 injuries, and 15.3 million dollars in direct property loss associated with Christmas tree fires in *residential* occupancies alone.[7]

■ NOTE

In 1990 alone, there were 29 deaths, 100 injuries, and 15.3 million dollars in direct property loss associated with Christmas tree fires in *residential* occupancies alone.

Each code has specific requirements. In general, only noncombustible or flame-resistant decorative materials may be used in assembly, educational, institutional, and mercantile occupancies and in hotels. The *IBC* and *BNBC* limits the amount of flame-resistant treated material to 10 percent of the total wall and ceiling area.

Where documented certification of flame resistance is not available for the inspector in the field, the field test prescribed by NFPA 701, *Standard Methods of Fire Tests for Flame-Resistant Textiles and Films*, should be conducted on a sample of the material. The result of failure to clearly understand the test procedures and criteria and to conduct the test in a safe environment are predictably unpredictable. Imagine trying to explain to the chief why the building fire occurred, or why your pants caught fire when the flaming, melted material dripped during the

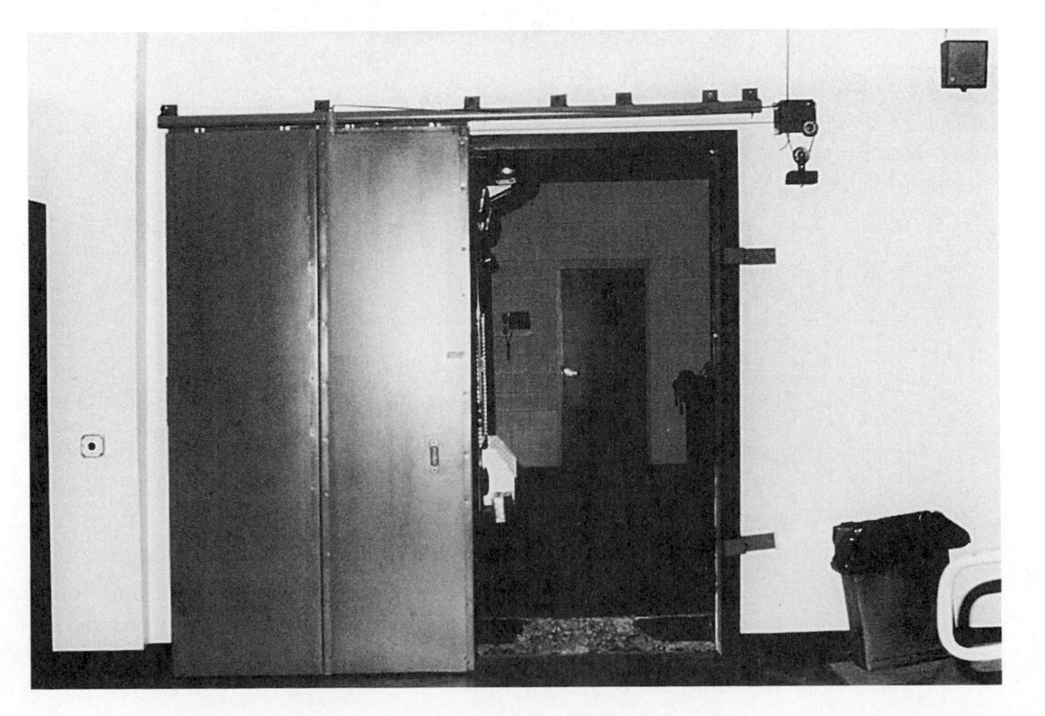

Figure 9-4 *Rolling fire doors must be tested annually by the owner.*

Table 9-2 *Interior Wall and Ceiling Finish Requirements by Occupancy[k].*

	SPRINKLERED[l]			UNSPRINKLERED		
GROUP	Vertical exits and exit passageways[a,b]	Exit access corridors and other exitways	Rooms and enclosed spaces[c]	Vertical exits and exit passageways[a,b]	Exit access corridors and other exitways	Rooms and enclosed spaces[c]
A-1, A-2	B	B	C	A	A[d]	B[e]
A-3[f], A-4, A-5	B	B	C	A	A[d]	C
B, E, M, R-1, R-4	B	C	C	A	B	C
F	C	C	C	B	C	C
H	B	B	C[g]	A	A	B
I-1	B	C	C	A	B	B
I-2	B	B	B[h,i]	A	A	B
I-3	A	A[j]	C	A	A	B
I-4	B	B	B[h,i]	A	A	B
R-2	C	C	C	B	B	C
R-3	C	C	C	C	C	C
S	C	C	C	B	B	C
U		No restrictions			No restrictions	

Notes: For SI, 1 inch = 25.4 mm, 1 square foot - 0.0929 m^2.

[a] Class C interior finish materials shall be permitted for wainscotting or paneling of not more than 1,000 square feet of applied surface area in the grade lobby where applied directly to a noncombustible base or over furring strips applied to a noncombustible base and fireblocked as required by Section 803.3.1 [of the source document].

[b] In vertical exits of buildings less than three stories in height of other than Group I-3, Class B interior finish for unsprinklered buildings and Class C interior finish for sprinklered buildings shall be permitted.

[c] Requirements for rooms and enclosed spaces shall be based upon spaces enclosed by partitions. Where a fire-resistance rating is required for structural elements, the enclosing partitions shall extend from the floor to the ceiling. Partitions that do not comply with this shall be considered enclosing spaces and the rooms or spaces on both sides shall be considered one. In determining the applicable requirements for rooms and enclosed spaces, the specific occupancy thereof shall be the governing factor regardless of the group classification of the building or structure.

[d] Lobby areas in A-1, A-2, and A-3 occupancies shall not be less than Class B materials.

[e] Class C interior finish materials shall be permitted in places of assembly with an occupant load of 300 persons or less.

[f] For churches and places of worship, wood used for ornamental purposes, trusses, paneling, or chancel furnishing shall be permitted.

[g] Class B material required where building exceeds two stories.

[h] Class C interior finish materials shall be permitted in administrative spaces.

[i] Class C interior finish materials shall be permitted in rooms with a capacity of four persons or less.

[j] Class B materials shall be permitted as wainscotting extending not more than 48 inches above the finished floor in exit access corridors.

[k] Finish materials as provided for in other sections of this code.

[l] Applies when the vertical exits, exit passageways, exit access corridors or exitways, or rooms and spaces are protected by a sprinkler system installed in accordance with Section 903.3.1.1 or Section 903.3.1.2 [of the source document].

Source: 2000 International fire Code®, Table 806.3. *Copyright 2000, International Code Council, Inc., Falls Church, Virginia. 2000 International Fire Code. Reprinted with permission of the author. All rights reserved.*

test. Know what you are doing and know the criteria for acceptance before you attempt the test in the field.

Flame resistant *does not mean that a material will not burn.* It means that if exposed to a small ignition source such as a cigarette lighter for a few seconds, the material burns little and self-extinguishes after the ignition source is removed. Treatments that render materials flame resistant must be renewed as needed.

Parade Floats

Parade floats are regulated by the *BNFPC* and *UFC.* All decorative materials used must be either noncombustible or flame resistant in accordance with NFPA 701. Motorized floats and tow vehicles must be equipped with fire extinguishers that are readily accessible to the operators.

UNSAFE STRUCTURES AND PROPERTIES

Vacant Structures

Vacant, dilapidated structures (see Figure 9-5) have a significant effect on the community. They become centers for illegal activity and vandalism and negatively impact the quality of the surrounding area. Property values drop, crime increases,

Figure 9-5
Unsecured vacant buildings can become shelters for the homeless and death traps for firefighters.

■ NOTE
Fire inspectors should be particularly vigilant in referring unsafe structures and in following up to see that action is taken; we have a vested personal interest.

people move out, and more properties are abandoned. Some would argue that the best thing that can happen in such instances is for the dilapidated buildings to burn down before the rest of the community is affected.

The model fire prevention codes term such buildings *unsafe structures* and require fire officials to refer such buildings to the building official for abatement, repair, or demolition. In many communities, building officials are limited to requiring owners to simply board up the first floor of such structures. Fire inspectors should be particularly vigilant in referring such structures and in following up to see that action is taken. We have a vested personal interest. It is the fire department that will be called on to extinguish fires in these structures.

Outdoor Waste Accumulation

The accumulation of wastepaper, wood, hay, weeds, or any combustible waste or rubbish *in quantities that constitute a fire hazard* (see Figure 9-6) are prohibited by the model fire prevention codes. In order to enforce these provisions, the fire inspector must remember that the intent of the model code groups in developing the fire prevention codes is to *prevent the spread of fire*, not to ensure that premises are neat in appearance, rodent free, or any of the other legitimate things that governments attempt to regulate.

The overgrown wood scrap pile in a backyard might be an eyesore for the neighbors. It might harbor rodents. It might be the very item that has caused a tiny

Figure 9-6 *"The trash service only comes once a week" is not an acceptable excuse for this unsafe condition.*

■ NOTE
Remember, you cannot make everybody happy; you can, however gain the public's respect by fairly enforcing the fire prevention code.

war within a neighborhood. But it *might not* be a violation. All that one of the participants needs is some reinforcements (like you), to "win" the battle. Many times the condition ceases to be the real issue. The battle becomes the issue.

Such instances require a cool, detached assessment of the situation, then decisive action on the part of the inspector. Do not allow yourself to become involved in a neighborhood dispute. Take the necessary action to gain compliance with the code. Violations of the code must be addressed. Remember, you cannot make everybody happy. You can, however, gain the public's respect by fairly enforcing the fire prevention code, referring matters outside your jurisdiction to the proper officials, and keeping all parties informed.

The relative hazard of overgrown vegetation varies from minor in the eastern United States, where rainfall is plentiful, to severe in many parts of the West. The *Uniform Fire Code* subsection 11.0.2.4, Combustible Vegetation, clearly underscores the hazard in the western states. The fire chief is authorized to order the removal of vegetation or the establishment of fire breaks in order to minimize the potential for brush fires.

ELECTRICAL HAZARDS

The installation of electrical services, wiring, fixtures, and permanent equipment is regulated by the electrical code official of the jurisdiction. NFPA 70, the *National Electrical Code*® (*NEC*®), is adopted throughout most of the United States. BOCA, SBCCI, and ICBO did not develop their own electrical codes. The ICC's *International Electrical Code* establishes administrative text necessary to administer and enforce NFPA's *National Electrical Code*®, or *NEC*®, and complies with electrical provisions contained in the other international codes.

Electrical safety is achieved through four general objectives. The fire prevention codes (1) require that permanent electrical systems be maintained as originally approved; (2) limit the use of temporary wiring; (3) prohibit the use of hazardous appliances, equipment, and wiring (see Figure 9-7); and (4) ensure access to electrical service equipment.

Temporary Wiring

The use of temporary wiring is limited to construction, renovation, and demolition projects, and for special events and holiday periods not to exceed 90 days. Outdoor Christmas decorative lighting, carnivals, and similar activities are legitimate uses of temporary lighting. Provisions for temporary wiring are included in the *NEC*®. In most jurisdictions, an electrical permit must be secured prior to the installation of temporary wiring, and the installation must be inspected. The fire inspector's job is to see that it was installed under permit and is maintained.

Extension cords, power strips, and multiplug adapters are encountered by inspectors on a daily basis. The fire prevention codes and the *NEC*® specifically address their use: Extension cords and power strips are only to be used to provide

Figure 9-7
This open electrical junction box is a violation and a hazard to the occupants of the dwelling.

power to portable appliances. Freezers, refrigerators, and the like are *not* portable appliances.

- Only grounded extension cords may be used for grounded appliances.
- The ampacity of the extension cord must equal or exceed the rated capacity of the appliance.
- Extension cords may not be affixed to wall or the ceiling, nor passed through walls, ceilings, or doorways nor under rugs or carpeting.
- Except for approved multiplug type cords, only one appliance may be powered by an extension cord.

Multiplug and cube-type plug-in adapters that do not comply with the *NEC* are specifically prohibited. Cube adapters, which enable building occupants to plug four or more appliances into a single receptacle, can easily lead to an overheating situation.

Hazardous Equipment

Appliances and equipment that are defective or have become unsafe due to lack of maintenance, misuse, or other damage must be removed from service. Fire inspectors should issue written orders and affix a tag prohibiting the use of the appliance or equipment. The electrical code official should be immediately notified.

Access to Electrical Service Equipment

Access to electrical service equipment is required for all occupancies. Clearance of at least 30 inches is required. Illumination is also required for the equipment service areas.

SMOKING

Smoking has become a national health issue, with significant political implications as well. The fire prevention codes address the fire safety aspect of smoking. Smoking and the improper disposal of smoking materials have played a significant role in our national fire experience for many years. The most common fatal fire cause in residential occupancies continues to be careless smoking.

In addition to the prohibitions against smoking contained within the fire prevention codes, many state and local governments prohibit smoking in business and mercantile occupancies, portions of restaurants, and other public buildings. Most of these ordinances are intended to protect the public from the nuisance and health hazards associated with secondhand smoke.

Citizen complaints regarding these ordinances are often directed to the fire prevention bureau. Informing an irate citizen that it is "not your problem," but you are really not sure who enforces the regulations is really bad. Trying to enforce regulations for which you have no legal authorization is worse. Citizen complaints should quickly be referred to the appropriate enforcing agency.

hazardous fire area term used in the *UFC* to describe public and privately owned areas of grass, brush, or forest with limited accessibility such that a fire originating in the area would be unusually difficult to extinguish and lead to significant damage and potential erosion

The fire prevention codes specifically prohibit smoking in areas where flammable liquids, explosives, and other similarly hazardous materials are stored, used, or handled. These provisions are found within the specific chapters or articles of the code. The fire official is also empowered to order the posting of no smoking signs and the designation of specific safe smoking areas where conditions warrant. Piers, wharves, and storage and industrial facilities all are likely candidates.

Smoking can also be prohibited during dry seasons in areas where there is significant potential for brush and wildland fires. The *UFC* has provisions for the fire chief to designate **hazardous fire areas**, where access as well as the use of any lighted material is prohibited without a permit. Smoking, firearms discharge, and other potentially hazardous activities are all prohibited or restricted as necessary during periods of extreme dryness.

NOTE

Where torches are used for soldering or removing paint, a capable fire watcher with an approved extinguisher must be posted.

OPEN FLAMES

Open flame devices such as heating and lighting appliances and torches for soldering pipe joints or removing paint are regulated within the general fire safety provisions of the fire prevention codes. Cutting and welding using acetylene is addressed within a specific chapter or article in each code.

The use of open flame devices in areas where readily combustible materials are stored, used, or handled is prohibited. Where torches are used for soldering or removing paint, a capable fire watcher with an approved extinguisher must be posted and remain on the scene for at least 30 minutes after the operation has been completed.

Heating appliances, vents, and chimneys must be installed and maintained in accordance with the building and mechanical codes and with their listings. The fire prevention codes require the sealing of unsafe appliances by the fire official. The use of portable fuel fired heating appliances is regulated by use group, depending upon the model code.

FIRE DEPARTMENT ACCESS

Emergency access on all-weather driving surfaces is required for every building. Not all buildings, however, are required to have marked fire lanes. Designated fire lanes are required for buildings by the *UFC* when any portion of the first floor is located more than 150 feet from the street.[8] Posted fire lanes (see Figure 9-8) are a

Figure 9-8 *Clear, posted fire lanes are critical to successful fire suppression and rescue operations. (Courtesy of Carl Maurice.)*

specific condition of increasing the allowable area of a building by the *BNBC*.[9] The *BNFPC* requires the fire official to designate fire lanes "as necessary" for the efficient and effective operation of fire apparatus."[10]

The *IFC, SFPC,* and *UFC* require fire lanes to be 20 feet in width, the *BNFPC* requires 18 feet. The minimum acceptable vertical clearance for modern fire apparatus is 13 feet 6 inches. Fire lanes should be conspicuously posted. Fire lane marking should be adopted as a local ordinance, or at the very least be a written directive from the fire official. Sign construction, size, color and size of lettering, as well as distance between signs and curb painting details should all be contained within a written procedure. Any deviation from the standard will result in future enforcement problems.

Summary

The model fire prevention codes prescribe general regulations that apply to all occupancies. These regulations are aimed at preventing ignition and limiting the impact of fire in the event of ignition. To prevent ignition, smoking and the use of open flames are regulated. Electrical equipment must be maintained and used in accordance with the listing. The impact of fires can be reduced through the maintenance of rated construction elements, by limiting combustible decorative material and combustible waste, and by providing adequate emergency access for the fire department.

Review Questions

1. List the two basic strategies used by the model fire prevention codes to safeguard life and property.
   1. ______
   2. ______
2. List three methods of preventing the ignition of fires by segregating the elements of the fire triangle.
   1. ______
   2. ______
   3. ______
3. List five methods of managing the impact of fire.
   1. ______
   2. ______
   3. ______
   4. ______
   5. ______
4. Which violation of the fire prevention code presents the largest potential for the number of safety deficiencies that can exist?
   ______
5. In dealing with an illegal change in use, the fire official's power is limited to what three actions?
   1. ______
   2. ______
   3. ______
6. List two barriers to fire that are not rated assemblies.
   1. ______
   2. ______
7. List two performance features of side-swinging fire doors.
   1. ______
   2. ______
8. Under what circumstances are hold-open devices permitted on side-swinging fire doors?
   ______
9. Where documented certification of flame resistance is not available for decorative material, what test procedure should be used by the inspector in the field? ______

10. List four general rules regarding the use of extension cords.
    1. ______
    2. ______
    3. ______
    4. ______

Discussion Question

1. During an inspection of a three-story office building, you note that mail slots have been installed in all the doors to the tenant suites. The doors open into the main exit stairway for the building and all are listed fire doors.
   a. Has the code been violated?
   b. What information would you request from the building owner?
   c. What is the potential significance of this situation?

Maintenance of Fire Protection Systems

Learning Objectives

Upon completion of this chapter, you should be able to:

- List the appropriate standards for the maintenance and inspection of sprinklers, standpipes, wet and dry chemical extinguishing systems, and fire alarm systems.
- Describe the importance of examining maintenance and inspection records kept by building owners.
- Describe what circumstances would cause an inspector to require an evaluation of an engineered fire protection system.
- Given flow test data, graph the available fire flow for a site.

Fire protection systems play an important role in our effort to protect the public through fire safety and building regulations. We recognize the potential effectiveness of these systems, which suppress, extinguish, or give early notification in fire situations. In many instances we reward building owners who install them by permitting increases in building height and area. We even reduce fire-resistance ratings based on the supposition that the fire protection systems will function as designed and installed, for the life of the building.

NOTE
The model building codes are designed to protect us all from our own actions, as well as those of our neighbors.

The model building codes are designed to protect us all from our own actions, as well as those of our neighbors. The fire wall between our town houses protects each of us from one another. If I am crazy enough to work on my motorcycle in the living room, the fire wall between us is the best friend you, as my next-door neighbor have. If the rating on the wall has been reduced because the developer installed residential sprinklers, the scenario changes a bit. You now need the fire separation wall *and my sprinkler system* to protect you from my unsafe actions. Expecting a guy who does major motorcycle repairs in his living room to maintain a sprinkler system might be taking a lot for granted.

NOTE
The effective maintenance of fire protection systems is a primary goal of every fire prevention inspections program.

With fire protection systems, all of our eggs are in the maintenance basket. Out of service or impaired systems can also give a false sense of security to the occupants, causing them to needlessly endanger themselves by attempting to extinguish a fire, or by not promptly evacuating. The effective maintenance of fire protection systems is a primary goal of every fire prevention inspections program.

INSTALLATION AND MAINTENANCE STANDARDS

All fire protection systems and equipment are installed under a standard prescribed by the building or fire prevention codes and are required by the fire prevention codes to be maintained. With the exception of water-based extinguishing systems, maintenance provisions for fire protection systems are included in the standards for installation, many of which are found in the following list:

NFPA 13 UFC 10-3	*Installation of Sprinkler Systems*
NFPA 13D UFC 10-4	*Installation of Sprinkler Systems in One- and Two Family Dwellings and Manufactured Homes*
NFPA 13R UFC 10-5	*Installation of Sprinkler Systems in Residential Occupancies Up to and Including Four Stories in Height*
NFPA 12	*Carbon Dioxide Extinguishing Systems*
NFPA 12A	*Halon 1301 Extinguishing Systems*
NFPA 17	*Dry Chemical Extinguishing Systems*
NFPA 17A	*Wet Chemical Extinguishing Systems*
NFPA 20	*Installation of Centrifugal Fire Pumps*

NFPA 24	*Installation of Private Fire Service Mains*
NFPA 72 UFC 10-2 UFC 10-3 UFC 10-4	*National Fire Alarm Code*
NFPA 14 UFC 10-6	*Standpipe and Hose Systems*
NFPA 10 UFC 10-1	*Portable Fire Extinguishers*
NFPA 25	*Inspection, Testing and Maintenance of Water-Based Fire Protection Systems*

NFPA 17, *Dry Chemical Extinguishing Systems* includes the installation and maintenance requirements for dry chemical systems. In 1995, NFPA 25, *Standard for the Inspection, Testing and Maintenance of Water-Based Fire Protection Systems* was issued. The standard incorporated requirements for all water-based extinguishing systems including private fire mains, fire pumps, and storage tanks. Detailed maintenance and inspection procedures as well as sample forms and documents are included in the standard.

WATER-BASED SYSTEMS

■ NOTE
An important part of every routine inspection is a review of all fire protection system records for the building.

Records of all fire protection system inspections, tests, maintenance, and repairs must be maintained on the premises by the owner or occupant. All test results, defects, and corrective actions must be logged. An important part of every routine inspection is a review of all fire protection system records for the building. Records should be well organized and up-to-date. They reflect the operational readiness of the fire protection systems in the building. The lives of the building occupants, as well as the firefighters who respond to an incident in the building, may depend on these systems.

Sample test and inspection report forms are included in NFPA 25. Appendix B in NFPA 25 contains thirty pages of sample checklists that cover the tests and inspections required for each water-based system. The forms were developed by the American Fire Sprinkler Association and National Fire Sprinkler Association and are available from those organizations.

■ NOTE
Closed valves are the primary cause of failure of water-based extinguishing systems in protected occupancies.

Sprinkler Systems

The inspection of every building equipped with a sprinkler system should include an inspection of the sprinkler riser and all control valves (see Figure 10-1). Closed valves are the primary cause of failure of water-based extinguishing systems in protected occupancies.[1] Begin at the riser where the water supply riser enters the

Figure 10-1 *Closed valves are the primary cause of sprinkler system failures.*

■ **NOTE**

The inspector should determine that the processes within the building are consistent with the occupancy hazard classification under which the sprinkler system was designed.

building. All sectional control valves should be visually inspected to ensure that they are open.

Note the hydraulic nameplate that is required to be attached to the riser on all hydraulically calculated systems. The nameplate must indicate the *design area, discharge density* (see Figure 10-2), *required flow and residual pressure demand at the base of the riser*, and *the hose stream demand*. As we discussed in Chapter 6, discharge density, or the amount of water discharged over a given area, is based on the occupancy hazard classification established in NFPA 13, *Standard for the Installation of Sprinkler Systems* (see Table 10-1). The inspector should determine that the processes within the building are consistent with the occupancy hazard classification under which the sprinkler system was designed.

If the hydraulic nameplate indicates that a system has been designed to discharge at a density of .15 gpm/sq. ft. over a 1,500 square foot area, the system has been designed at the Ordinary Group 1 curve. If the building is being used as a printing facility, the occupancy hazard classification from Table 10-1 has obviously changed, and our system must be reevaluated for the greater hazard. It must now be recalculated at the Ordinary Group 2 curve. A notice of violation should be issued requiring an evaluation of the system, and, if indicated, requiring an upgrade of the sprinkler system to protect the current hazard.

While in the sprinkler room, the stock of spare heads should be inspected and the heating appliances should be checked. Dry pipe valves must be main-

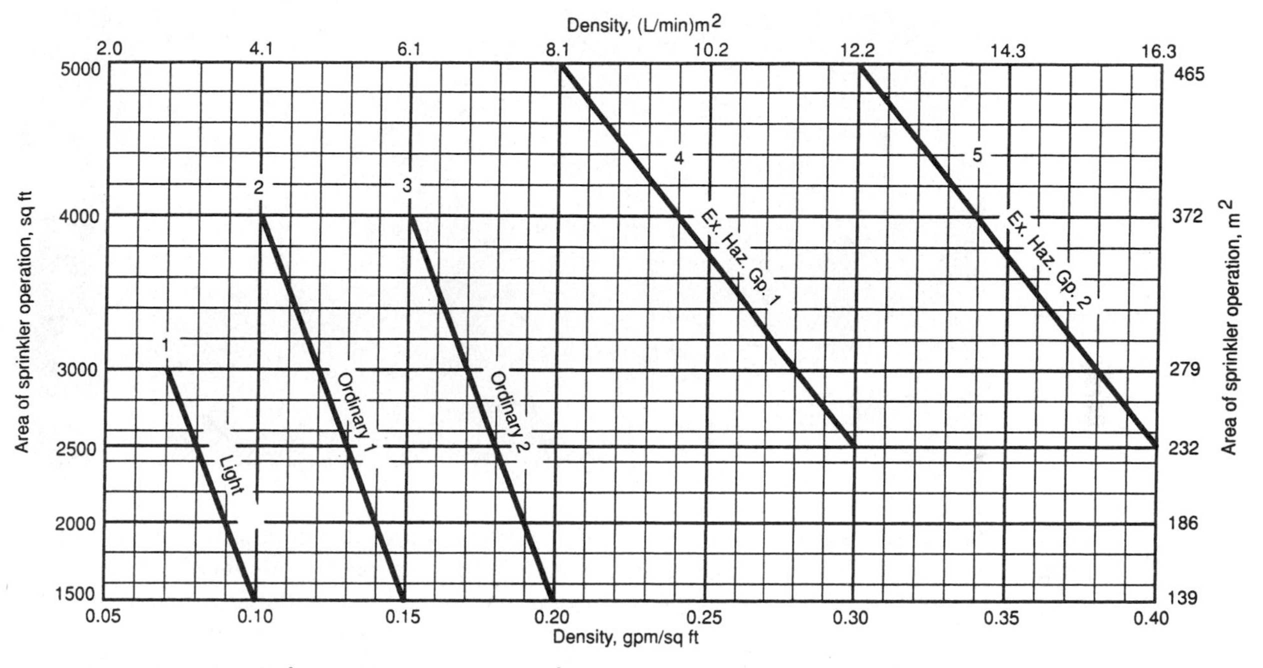

For SI Units: 1 sq ft = 0.0929 m^2; 1 gpm/sq ft = 40.746 (L/min)/m^2.

Figure 10-2 *Sprinkler density is dependent on occupancy classification. (Reprinted with permission from NFPA 13,* Installation of Sprinkler Systems, *Copyright © 1999, National Fire Protection Association, Quincy, MA 02269. This reprinted material is not the complete and official position of the National Fire Protection Association on the referenced subject which is represented only by the standard in its entirety.)*

Table 10-1 *Occupancy classification examples from NFPA 13.*

Classification	Examples
Light Hazard	Churches, hospitals, schools, offices, theaters, residences
Ordinary Group 1	Parking garages, canneries, laundries, restaurant service areas, electronic plants
Ordinary Group 2	Dry cleaners, horse stables, machine shops, post offices, print shops, library stack rooms
Extra Hazard Group 1	Aircraft hangars, sawmills, plywood manufacturing
Extra Hazard Group 2	Flammable liquid spraying, plastic processing, varnish and paint dipping

Figure 10-3
Improperly placed insulation can lead to frozen sprinkler pipes.

NOTE
Water damage from frozen pipes can cause entire communities to develop strong feelings against sprinklers.

tained at a minimum 40°F. Wet pipe systems must also be maintained at a minimum 40°F. Frozen pipes (see Figure 10-3) unnecessarily endanger building occupants by voiding the sprinkler protection and do considerable water damage. Water damage from frozen pipes can cause entire communities to develop strong feelings against sprinklers. The inspection of insulation, heating appliances, and methods of preventing trapped water in dry piping are every bit as important as checking the supply valves.

NFPA 25 includes provisions for visual inspection of components, the test and manipulation of valves and controls, the removal and test of antifreeze solutions, and the removal and laboratory test of heads according to a specific schedule. Corporations with effective safety and risk management programs and those whose insurance companies retain firms such as Factory Mutual Engineering often meet or exceed the requirements for maintenance and testing. Most others only recognize the importance of sprinkler testing and maintenance with the assistance of the fire official.

Standpipe Systems

Visual inspection and testing are required by NFPA 25 for standpipe and hose system components. Five-year flow tests in which water is flowed from the highest or most remote point in the system should never be waived due to inconvenience. Flow tests should not be conducted during freezing weather due to the obvious safety concerns, but should be conducted when temperatures moderate. Flow tests are the most effective means of determining a continued adequate water supply and of effectively testing the entire supply system from the street to the topmost outlet. An effective inspection system *will* identify water supply problems caused by obstructions and such problems as partially closed valves in the municipal water system. Closed valves, foreign material, and even water mains that

■ **NOTE**
Flow tests are the most effective means of determining a continued adequate water supply and of effectively testing the entire supply system from the street to the topmost outlet.

have been crushed by heavy equipment may lurk beneath the asphalt and concrete. Water supply problems are not uncommon occurrences. They must be detected and promptly resolved.

Other components that require testing and inspection include pressure regulating devices, which must be tested every 5 years to ensure that adequate flow and pressure is provided at each hose outlet. Standpipe hose and nozzles must also be tested.

Fire Pumps

Fire pumps might be considered the most critical element in many water-based extinguishing systems. While all portions of the system are important, a lot more can go wrong with a fire pump than with piping, valves, and tanks. The lack of maintenance or misuse can lead to expensive repairs that some owners would just as soon forego at the risk of disabling the standpipe and sprinkler systems. Fire pumps must be run and inspected weekly and subjected to an annual flow test. Your inspection of the fire pump includes an inspection of the building engineer's records and his knowledge of the fire pump. Time spent educating building engineers and assisting them in complying with the code is time well spent.

Other water-based systems and components such as fixed water spray and foam as well as water tanks, private fire service mains, and fire department connections (see Figure 10-4) have fixed maintenance and inspection schedules specified by NFPA 25.

Figure 10-4 *Fire department connections must be visible and accessible, even during construction and renovation.*

■ NOTE

Whether the fire department actually conducts flow tests or merely reviews the test data provided by another organization, knowledge of proper test practices and procedures and the ability to "understand the picture" when test data is presented in graphic form are fundamental skills for the fire inspector.

FIRE FLOW TESTS

Flow testing is the most efficient and effective method of verifying an adequate water supply for fixed fire suppression systems and of ensuring adequate water supply for manual fire extinguishment by the fire department. Unfortunately, it is a function that has, in large part, gone by the wayside due to the constantly expanding list of responsibilities that many modern fire departments have undertaken. Whether the fire department actually conducts flow tests or merely reviews the test data provided by another organization, knowledge of proper test practices and procedures and the ability to "understand the picture" when test data is presented in graphic form are fundamental skills for the fire inspector.

The term *flow testing* includes routine tests from main drains and inspector's test pipes, flows from the topmost or most remote standpipe outlets, fire pump flow tests, and hydrant and water supply system tests. There are three major points to keep in mind about flow tests:

1. The frequencies for testing contained within the model codes and referenced standards are minimum frequencies *not* optimal frequencies. If they are not being performed at least at the required frequency, trouble could be brewing.
2. Testing accomplishes more than just determining that the same amount of water flows through the discharge orifice from one test to the next. Tests provide sorely needed flushing of sediment and foreign material from mains and piping.
3. Testing of municipal water systems is the best way for firefighters to learn about the municipal water system, and where the best flows and the worst flows are available. This knowledge can be of vital importance. In the early stages of major incidents, the road system around major incidents quickly becomes blocked by vehicles, making efficient hose lays almost impossible after initial companies have been positioned. Adequate water supply may depend on a connection to the municipal water system at a point with adequate main size, not necessarily the main that is adjacent to the fire incident.

The NFPA 1031 (*Professional Qualifications for Fire Inspector*) committee recognized the trend away from hands-on flow testing in the fire service, which is reflected in the Job Performance Requirement (JPR) for Fire Inspector II (1999):

> Verify fire flows for a site, given fire flow test results and water supply data, so that required fire flows are in accordance with applicable codes and standards and all deficiencies are identified, documented, and reported in accordance with the policies of the jurisdiction.

Whether your department actually performs flow tests or not, you need to be familiar with proper test procedures to be able to evaluate test data submitted by other organizations. NFPA 291, *Recommended Practice for Fire Flow Testing and Marking of Hydrants*, should be consulted for guidance.

Flow Test Considerations

First and foremost, flow tests must be conducted safely. Care should be taken to prevent property damage and to minimize the impact on adjacent businesses and residences. Flow tests must be coordinated with the local water authority. Property damage ranging from the interruption of a manufacturing process to ruining a family's wash with rust and scale can be avoided with a little planning.

Always perform a safety check prior to conducting a flow test. The safety of the public and of fire department personnel must be your first priority. Consider these items:

- Weather conditions: Will the water drain to a safe location prior to freezing? Creating icing conditions on streets, sidewalks, or parking lots is simply not an acceptable result of a routine testing procedure, conducted in the name of safety.
- Time of day: Will the test disrupt traffic during rush hour or create a traffic hazard? Will the test negatively impact business operations or cause hardship to the occupants of adjacent residences? Could children walking to school be impacted?
- Topography: Where will the water drain? Are there open manholes or underground vaults, construction operations, or utility workers in the vicinity? Could runoff endanger workers or damage equipment?
- Location: Are hydrants situated so that flowing water strikes automobiles or pedestrians, damages landscaping, or undermines property?

The potential benefits from conducting flow tests are enormous. The potential for damage, injury, or a public relations nightmare is real. Proper planning and coordination will minimize the possibility of a problem.

The potential benefits from conducting flow tests are enormous; the potential for damage, injury, or a public relations nightmare is real.

Flow Test Equipment

In order to conduct a hydrant flow test, the following equipment is needed (see Figure 10-5):

1. A good quality (preferably steel) ruler with 1⁄16th inch divisions
2. A pitot tube with a bourdon gauge
3. A 2½-inch hydrant cap, with a bourdon pressure gauge and petcock blow-off
4. A hydrant wrench(s)
5. Writing paper and pencil to record test data and sketch the area

Flow Test Procedures

The test hydrant is selected due to its proximity to a particular building or process. The test procedure is designed to determine the available flow from the water sys-

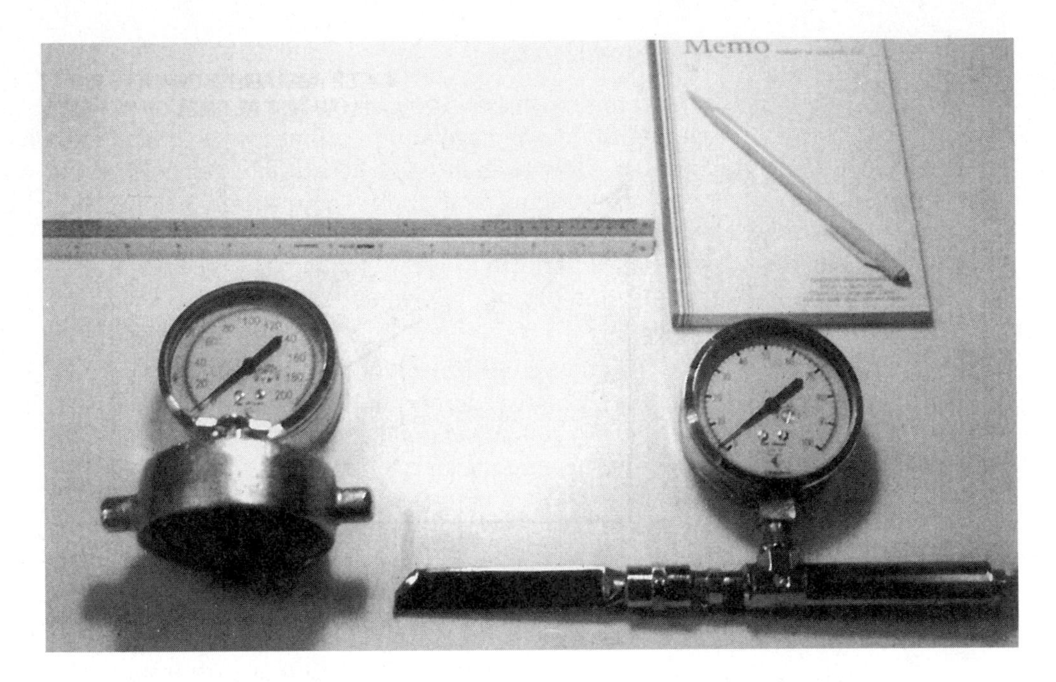

Figure 10-5 *Water flow test equipment. Clockwise from top left: ruler to measure hydrant opening, notebook to record flow test results, pitot tube and bourdon gauge, hydrant cap with bourdon gauge.*

tem at the test hydrant. Flowing of the test hydrant is limited to flushing it to remove sediment and debris and to ensure adequate pressure readings. A flow hydrant is downstream for the water source and test hydrant and should be selected to minimize potential property damage and traffic disruption. Refer to a map of the water system or check with the water authority to determine direction of flow within the water system.

When operating fire hydrants, open and close them slowly to prevent water hammer. Always open them fully, then, if necessary, throttle down by closing several turns. Note for proper drainage in dry hydrants. Blocked drains will lead to frozen hydrants and must be immediately reported to ensure their timely repair. Always check for damage to threads, obstructions such as shrubbery or fences, and damage. Report these conditions to the proper authority.

At the test area, after making the appropriate notifications and safety check, the following sequence should be followed:

static pressure
pressure exerted within a water system at no flow

1. Fully open and flush the test hydrant until it flows clear. Failure to flush the test hydrant may introduce sediment or debris into the gauge. Attach the 2½-inch cap with pressure gauge. Fully open the test hydrant, open the petcock to bleed off air from the barrel, and close the petcock. Note and record the pressure reading. This is the **static pressure**, or pressure exerted within the water system at no flow. (See Figure 10-6.)
2. Remove a 2½-inch cap from flow hydrant number one (additional flow hydrants may be required). Measure and record the inside diameter of the discharge (Figure 10-7). Reach into the discharge (Figure 10-8) and determine its

Figure 10-6 *Flow test procedure at test hydrant: Record the static pressure before opening the flow hydrant and then record the residual pressure after opening the flow hydrant. (Courtesy Superior Automatic Sprinkler Corp.)*

Figure 10-7 *Flow test procedure at flow hydrant: Accurately measure the inside opening to the nearest 1/16 inch. Record this value. (Courtesy Superior Automatic Sprinkler Corp.)*

Figure 10-8
Checking the hydrant coefficient at the flow hydrant: Determine the hydrant coefficient (C_d) (rounded, square, or projected) by referring to Figure 10-9. Record this value. (Courtesy Superior Automatic Sprinkler Corp.)

hydrant coefficient
a number describing the opening from which water flows from a hydrant

pitot pressure
the pressure (in psi) shown on the bourdon gauge attached to the pitot tube inserted in a water stream flowing from a fire hydrant

residual pressure
the pressure (in psi) measured at the test hydrant with water flowing from the flow hydrant

shape. Then choose the **hydrant coefficient** of discharge, a number that describes the opening, as shown on Figure 10-9, and record it.

3. Fully open the flow hydrant until it runs clear. Insert the pitot tube into the stream as shown on Figure 10-10. Note and record the pressure indicated on the bourdon gauge attached to the pitot tube; this is referred to as **pitot pressure**. It may take a few minutes for the pressure to stabilize. If readings fluctuate, use the average pressure.
4. While water is flowing from the flow hydrant(s), note and record the pressure indicated by the gauge on the test hydrant. This is the **residual pressure**. Additional downstream hydrants may have to be flowed simultaneously in order to achieve at least a 10-psi drop in pressure at the test hydrant.
5. Once satisfactory test information has been obtained, the flow hydrants can be slowly shut down and checked for proper draining. Any abnormality should be reported to the water authority.
6. When shutting down the test hydrant, the petcock on the 2½-inch gauge should be opened. Failure to open the petcock allows water draining from the barrel to create a partial vacuum, potentially damaging the gauge. Ensure that the flow hydrant drains.

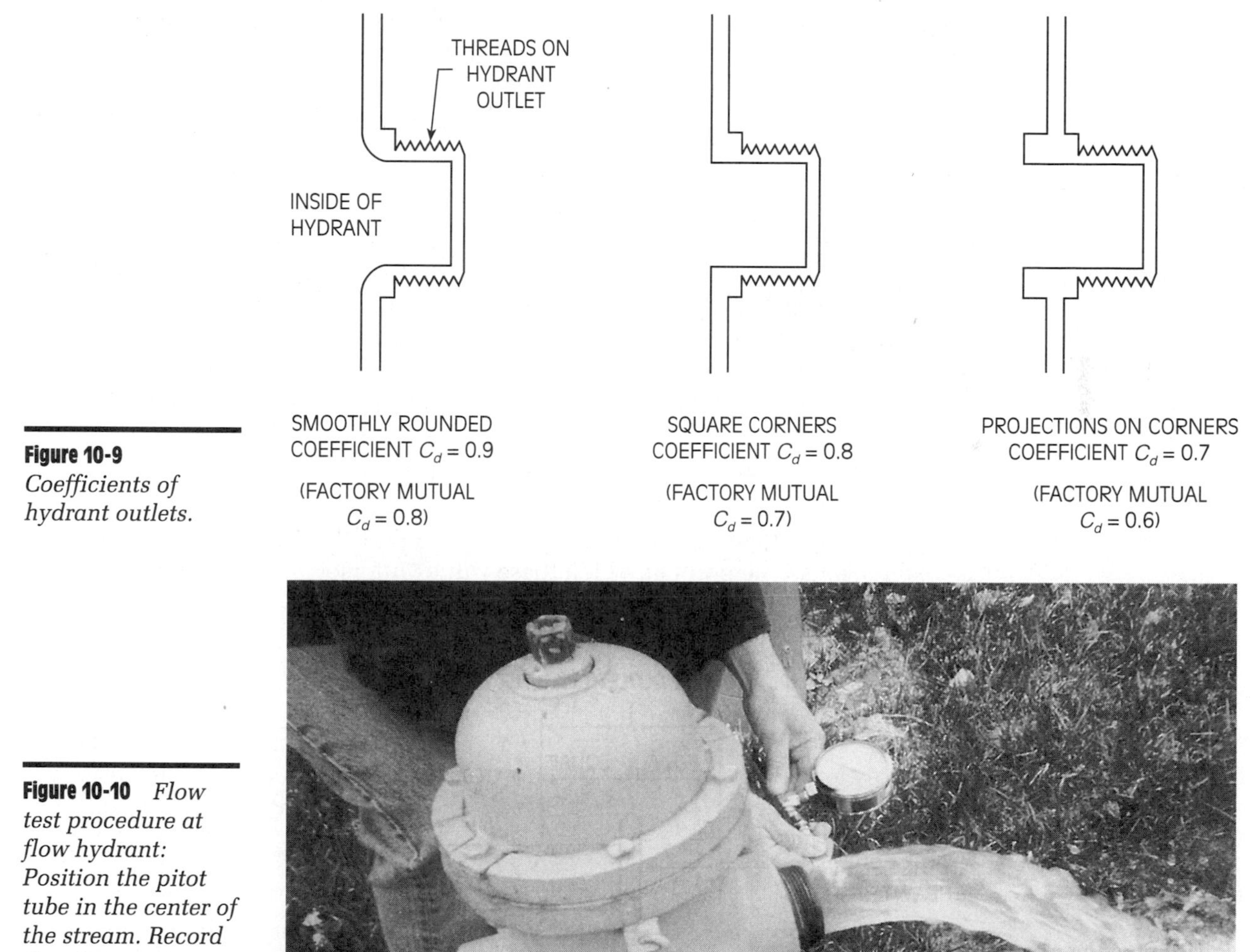

Figure 10-9 *Coefficients of hydrant outlets.*

Figure 10-10 *Flow test procedure at flow hydrant: Position the pitot tube in the center of the stream. Record the pitot pressure reading displayed on the bourdon gauge. (Courtesy Superior Automatic Sprinkler Corp.)*

Graphing Fire Flow Data

Information from flow tests can be plotted using logarithmic graph paper, entered into a computer program, or calculated using flow tables or formulas. To convert pitot pressure to flow, use the following formula:

$$\text{Flow (in gpm)} = 29.83 \times D^2 \times C_d \times \sqrt{P}$$

where D is the inside diameter of the discharge orifice, C_d is the coefficient of discharge for the hydrant type, and P is the pitot pressure.

Graphing flow data using logarithmic graph paper is a quick method to compare flows with those of previous years. All that is needed is flow test data, 1.85 exponential, or logarithmic, graph paper, a straight-edge, and a very sharp pencil. (Mechanical pencils are extremely useful.)

Logarithmic graph paper printed specifically for flow tests such as shown in Figure 10-11 is available from the National Fire Sprinkler Association and engineering supply companies. The *x*-axis (horizontal) represents flow; the *y*-axis represents pressure. Note that there are several scales for the *x*-axis to accommodate a wide range of flows. To compare test results with those of previous tests, it is important to use the same scale.

Figure 10-11 is an example of graphing fire flow data using the following steps:

1. Locate and mark the static pressure on the *y*-axis (vertical) at 0 gpm flow (point 1).
2. Locate the total test flow along the *x*-axis and the residual pressure along the *y*-axis. The point at which these values intersect (point 2) represents the residual pressure reading at the test hydrant at total flow from your flow hydrant(s). Mark this point.
3. Use a straightedge to connect the two points, extending the line from the *y*-axis base line to the *x*-axis base line.
4. Locate the point at which the line you have just drawn intersects the 20-psi line along the *y*-axis (point 3.) This point on the *x*-axis is your total available flow at 20 psi.

The Necessity of Flow Testing

Fire flow from hydrants is by definition "the amount of water that is available at 20 psi residual pressure for firefighting." There is actually more water available, but operating fire department pumpers at lower residual pressures result in a negative pressure at points within street mains. This negative pressure could cause the collapse of the mains or other water system components, or back-siphonage of polluted water from some other interconnected sources. Operating at residual pressures of less than 20 psi (1.4 bar) is not permitted by many state health departments.

Current results should be compared with previous tests. Changes may reflect increased usage, the addition of equipment or capacity by the water authority, or obstructions such as foreign material or partially closed valves. Changes should be reported to the water authority for investigation and resolution. Valves that have been closed due to construction or maintenance and then not fully opened are a common but significant impairment that may only be identified by flow testing or failure of the system during a fire incident.

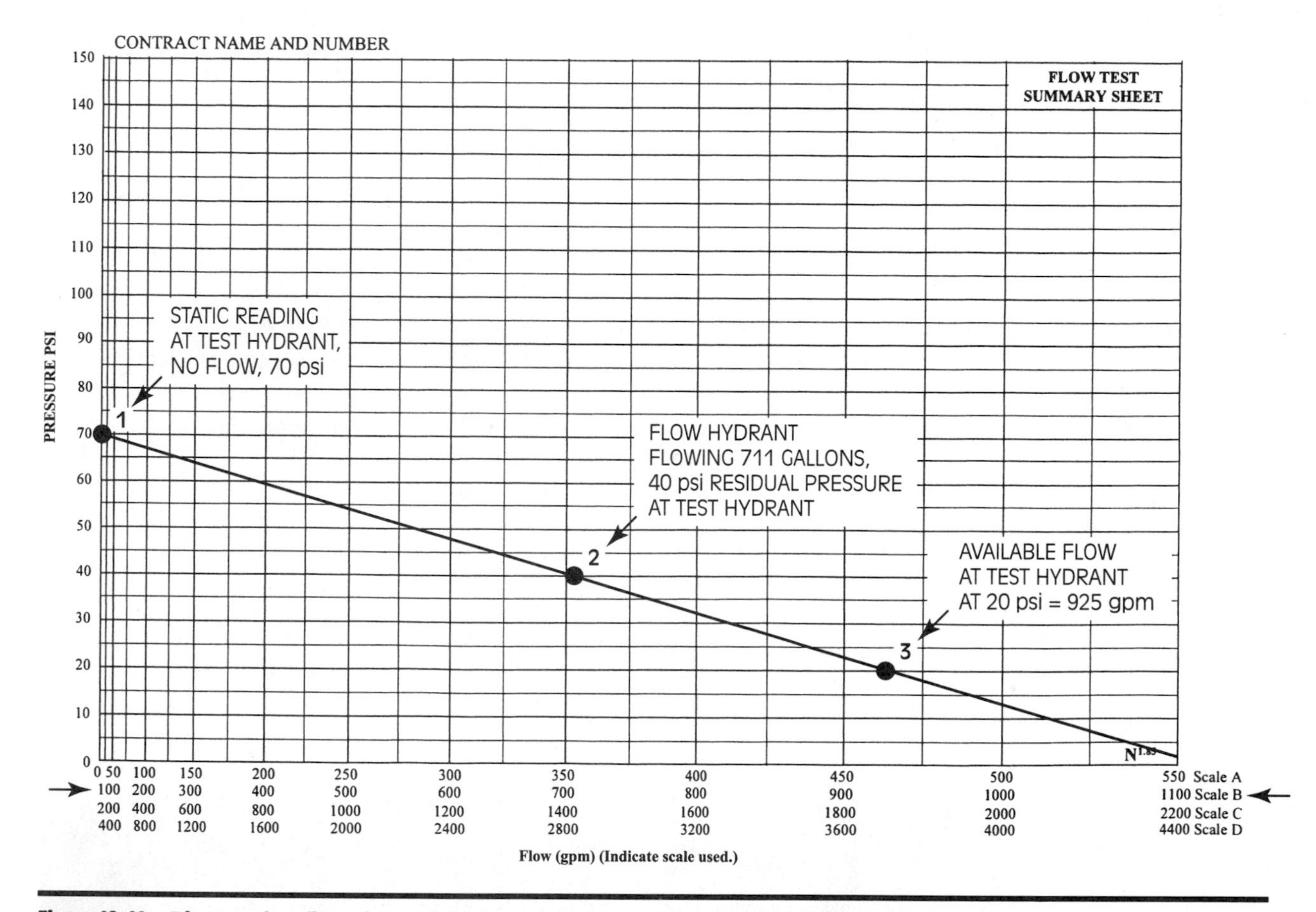

Figure 10-11 *Plotting fire flow data. In this example, the "B" scale was chosen. Point 1 reflects the static pressure reading at the test hydrant; 2 reflects the pressure reading at the test hydrant while the flow hydrant was operating; and 3 is the point at which the flow curve intersects the 20 psi line, indicating that flow is available at 20 psi.*

WET AND DRY CHEMICAL EXTINGUISHING SYSTEMS

Wet and dry chemical extinguishing systems are used to protect various industrial hazards from commercial cooking operations to spray-painting booths. The systems must be maintained in accordance with their respective standards (NFPA 17A and 17), the manufacturers' recommendations, and the fire prevention codes.

The *IFC, BNFPC,* and *UFC* require testing of all operating components every 6 months. Key items for the inspector to check are:

- Adequate system pressure
- Adequate coverage (appliances or process has not changed)
- Buildup of grease (see Figure 10-12) or residue on fusible links or elements
- Obstructions to mechanical operation of the system
- Obstructions to manual activation controls (see Figure 10-13)
- Placement of required supplementary fire extinguishers

NOTE
Any mixing of extinguishing agents should be strictly avoided.

In Chapter 6 we discussed the four classes of fire and some of the different agents used to extinguish them. Any mixing of extinguishing agents should be strictly avoided. Although some agents such as potassium bicarbonate dry chemical and aqueous film-forming foam have been found to be effective together, most are not. The different substances can react and reduce or even eliminate the effectiveness of one or both. A good example is the incompatibility between the two dry chemical agents potassium bicarbonate (BC) and ammonium phosphate (ABC).

Effective June 30, 1998, NFPA 10, *Standard for Portable Fire Extinguishers,* has required the installation of a fire extinguisher with a Class K rating as supplemental protection for commercial cooking operations. The standard specifically excepts "extinguishers installed specifically for these hazards prior to June 30,

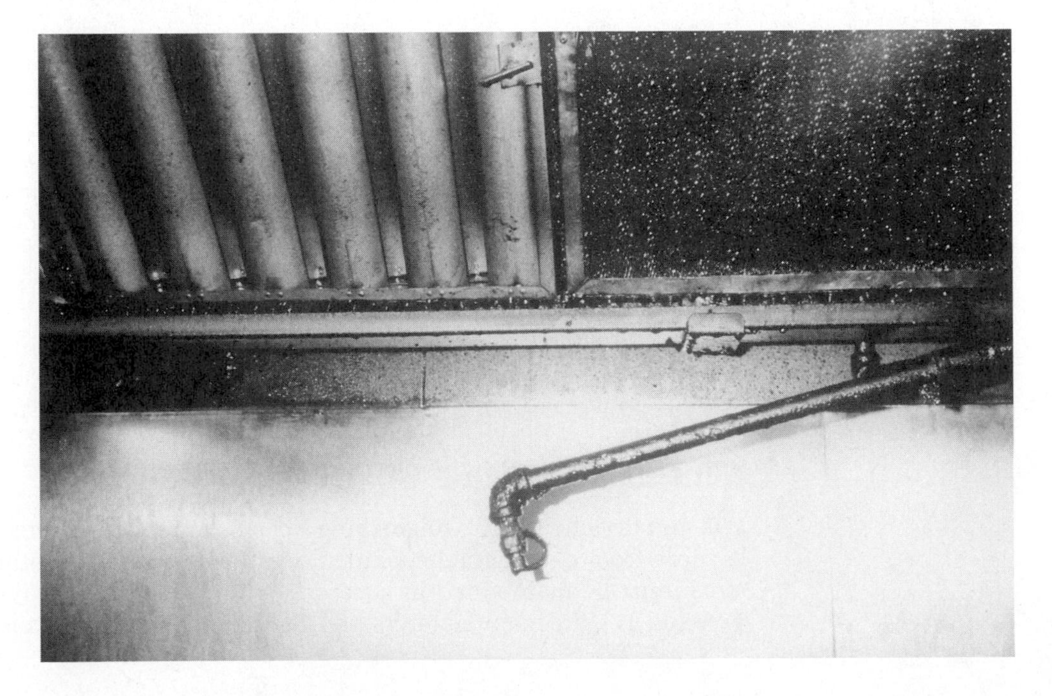

Figure 10-12 *This system is filthy, and it has an unapproved mesh filter and unsupported piping. The inspector should order cooking operations under the hood discontinued until corrections are made. (Courtesy of Ron Berry.)*

Figure 10-13 *Manual activation of the fire suppression system is blocked by kitchen equipment. (Courtesy of Ron Berry.)*

1998." (NFPA 10, 2-3.2) Any system installed or upgraded after June 30, 1998, must meet the new requirement. Fire equipment manufacturers are no longer manufacturing replacement parts for dry chemical, commercial cooking fire protection systems. The lack of available parts for the upgrade of systems to meet the UL 300 Standard, and, at the same time, require the installation of Class K extinguishers.

NOTE
The addition of cooking appliances or even the upgrade of an appliance may tax a system beyond its design capabilities.

Inspectors must be on the alert for changes in the hazards that these engineered systems have been approved to protect. The addition of cooking appliances or even the upgrade of an appliance may tax a system beyond its design capabilities. There is very little fat in engineered fire protection systems. The replacement of a commercial fryer with a modern high efficiency unit that has multiple heating elements and 50 percent more cooking oil as a fuel load requires an engineering evaluation of the fire suppression system.

HALON AND CLEAN AGENT SYSTEMS

NFPA Standards 12A and 2001, respectively, include installation and maintenance provisions for total flooding halon and clean agent fire extinguishing systems. System components (detection and initiating devices) must be inspected and tested at regular intervals. *This testing does not involve discharging the systems!*

FIRE ALARM SYSTEMS

In 1993 NFPA consolidated the various fire alarm and detection equipment standards into NFPA 72, *The National Fire Alarm Code.* The installation, inspection, testing, and maintenance of all fire alarm systems and equipment are included in NFPA 72. An effective program of enforcing the maintenance and inspection provisions for fire alarm systems will result in a decrease in responses to nuisance alarms by the fire department. Reducing the number of nuisance alarms not only saves time and money, but potentially saves lives as well. The lives of the public are better protected because companies spend less time out of service on nuisance alarms, and the firefighters and public face less exposure to the traffic hazards associated with emergency response.

■ NOTE

Reducing the number of nuisance alarms not only saves time and money, but potentially saves lives as well.

Summary

With few exceptions, the fire protection systems that provide building and life safety protection are engineered; that is, they are designed for a specific hazard. Standards for installation, maintenance, testing, and inspection are referenced in the model building and fire prevention codes. The job of the fire inspector is to ensure that the system is functional, that the maintenance and inspection provisions are being followed, and that the hazard the system was designed to protect has not changed.

Review Questions

1. What standard was developed for the maintenance of water-based extinguishing systems by the National Fire Protection Association?

   ____

2. List three items of information that should be included in fire protection system records.
   1. ____________________
   2. ____________________
   3. ____________________

3. What is the primary cause of failures in water-based extinguishing systems?

   ____

4. List four items that must be included on sprinkler system hydraulic nameplates.
   1. ____________________
   2. ____________________
   3. ____________________
   4. ____________________

5. List the five occupancy classifications from NFPA 13.
   1. ____________________
   2. ____________________
   3. ____________________
   4. ____________________
   5. ____________________

6. Wet pipe sprinkler systems must be maintained at a minimum of ____ degrees to protect against freezing.

7. What is the most effective method of determining an adequate water supply for an existing sprinkler system?

   ____

8. List six key items to be checked during an inspection of a wet or dry chemical extinguishing system.
   1. ____________________
   2. ____________________
   3. ____________________
   4. ____________________
   5. ____________________
   6. ____________________

9. Supplementary fire extinguishers installed in the vicinity of range hood fire suppression systems should be rated for Class ____ only.

10. What standard for the installation and maintenance of fire alarm systems was developed by the National Fire Protection Association?

    ____

Discussion Question

1. During an inspection of a restaurant, you notice that a new range and fryers have been installed. The old appliances had been in place since the establishment was first opened over 20 years ago. The restaurant owner informs you that the equipment salesman said they did not need a permit for installation because it was "replacement with like equipment."
   - a. Does a code violation exist?
   - b. What action should the fire inspector take in this case?
   - c. Explain your actions to the chief who receives a call from a city councilman who claims you are harassing the restaurant owner.

Means of Egress Maintenance for Occupancies

Learning Objectives

Upon completion of this chapter, you should be able to:

- Name five ways that the means of egress from a building or space can be compromised.
- Describe the importance of posted occupant loads in assembly buildings.
- Describe the process used to establish the occupant load.
- Describe what constitutes "overcrowding" in an assembly occupancy.

Throughout this text, we have discussed the model codes' balancing act with respect to building safety elements used in building design. Requirements for fire-resistive construction features, fire protection, interior finish, and means of egress are based on the needs of the building occupants and on the relative hazards associated with the building use. In theory, at least, when buildings are designed and constructed in accordance with model codes, every element of safety and security is in perfect harmony. People are safe, happy, and productive.

Unfortunately, that just is not the case. From the time the certificate of occupancy is issued by the building official to the time the building is mercifully torn (or unmercifully burns) down, it is under a constant assault by the owners, occupants, and outsiders. In the name of doing things better, the balance created by the model codes is often lost. Unquestionably, of all the elements of building safety prescribed by the model building codes, *means of egress* is the easiest to compromise. It generally takes time and money to seriously affect other major building features. It only takes a conscientious sales clerk to realize that all that valuable space in the back hallway and stairwell is going to waste before the big Christmas sale. She dutifully fills the exit corridor and stairs with combustible storage. The overheated dishwasher who props open the door leading into the exit corridor of the shopping mall is only attempting to get some ventilation. In cooling off the kitchen he puts hundreds of people at risk from a fire incident anywhere within the mall. Simple, everyday occurrences that compromise the means of egress combine with the not so common, to threaten the occupants in the event of fire.

MAINTAINING MEANS OF EGRESS

■ NOTE
The length of travel in the exit access is regulated, which is not the case in the exit or the exit discharge.

■ NOTE
Every component of every means of egress must be checked by the inspector.

The same elements involved in the original establishment of the means of egress must be revisited in order to assess the adequacy of the exits from any building or space. First remember the three parts of a means of egress (see Figure 11-1): *exit access*, *exit*, and *exit discharge*. Not all parts are regulated in the same fashion. The length of travel in the exit access is regulated, which is not the case in the exit or the exit discharge. Storage is not permitted within an exit, but is acceptable within most parts of the exit access.

In Chapter 7 we saw that a seven-step process can be used to design or assess the adequacy of the means of egress from a building or space. We used many of the same steps, in a bit different order, to determine the capacity of the different means of egress components in existing spaces. Fortunately, these exercises are not necessary on most inspections. Most of the things people do that compromise the means of egress fall into five broad categories. Exits are obstructed or, obscured, the integrity of their construction is compromised, the integrity of their function is compromised, or the exit is eliminated. In order to perform an adequate inspection of any building or space *every component of every means of egress* must be checked by the inspector. The inspector must walk every stair and corridor and open every door. Without a methodical check of each component, the inspection is incomplete. The means of egress from a high-rise building with forty stories of

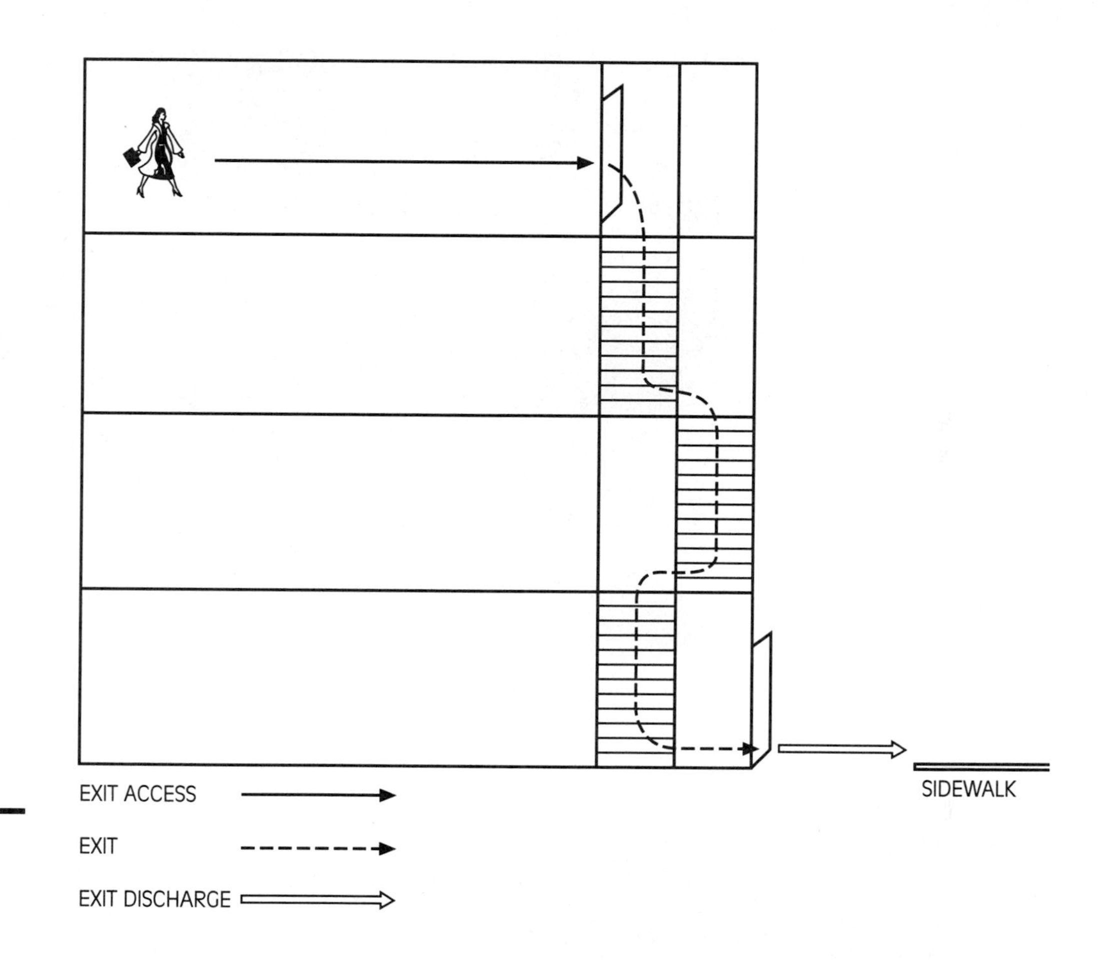

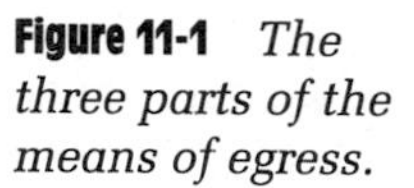

Figure 11-1 *The three parts of the means of egress.*

perfectly operating fire doors is worthless if the stairway discharge door at the ground floor is blocked from the outside by shrubbery and cannot be opened.

OBSTRUCTIONS

■ **NOTE** **By far the most common violation regarding the means of egress is the blocking of doors, aisles, or passageways.**

By far the most common violation regarding the means of egress is the blocking of doors, aisles, or passageways (see Figure 11-2). Often the condition is remedied by the occupants within minutes of occurrence. Other times the obstruction is permanent, or semipermanent. Restaurants commonly store high chairs or stacks of extra chairs in passageways that lead to rear exits, making them impassable. The same passageway, which probably also leads to the rest rooms, is a great place for the cigarette machine. Reducing the width of the passageway by the width of the cigarette machine probably creates a means of egress that falls below the code-required minimum width.

Figure 11-2 *Exits can become blocked during deliveries.*

■ **NOTE**

Shopping malls and regional shopping centers should be frequently checked prior to and during the holiday shopping season.

Obvious obstructions such as storage are generally not difficult for the inspector to remedy. Most building owners and occupants can understand that doors and aisles should not be blocked. For them it is a matter of degree: They usually feel that they did not have it blocked *that much* and *people could have gotten out if they really needed to.*

Unfortunately, sometimes reason does not prevail. Some business owners insist on obstructing the means of egress, sometimes in order to accommodate additional stock and at other times for sheer convenience. Instances where large retail chains have repeatedly obstructed the means of egress during the holiday shopping season are numerous. Shopping malls and regional shopping centers should be frequently checked prior to and during the holiday shopping season with advice to management that all mercantile occupancies are being similarly inspected and to expect them to be checked again.

Obstructions such as the cigarette machine or other equipment (see Figure 11-3) in the rear corridor can be much more difficult. Small business owners often have few options due to limited space. "It has been like this for years and none of the other inspectors ever said anything," is also a common theme. If the cigarette machine reduces the width of the passageway below the code-required minimum, you must order the owner to move it to another location.

Egress ways can also be obstructed without the knowledge or assistance of the owner or tenant. Snow and ice that have accumulated on exterior stairs, fire

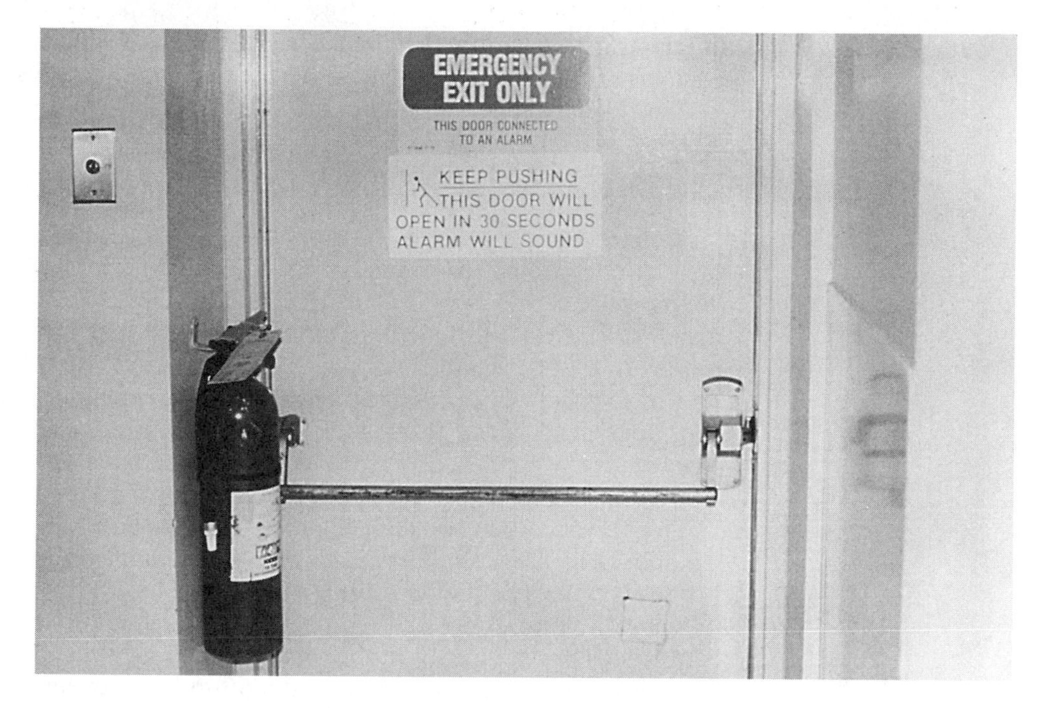

Figure 11-3 *This fire extinguisher reduces the clear opening and is a hazard. (Courtesy of Duane Perry.)*

■ **NOTE**

Snow and ice that have accumulated on exterior stairs, fire escapes, or in front of exterior doors can reduce or even eliminate entire exits.

escapes, or in front of exterior doors can reduce or even eliminate entire exits. Trees or shrubs have been known to grow to such proportions that exterior doors are blocked, again without the knowledge of the building occupants.

OBSCURATION

Means of egress components are generally obscured by one of two methods. In the first, for whatever reason, the exit is hidden from view. Draperies used for privacy can camouflage doors, making them unavailable for the public. Large bookcases or display cases can also be arranged in ways that may not obstruct access to the exit, but which block any view of the exit making it just as unusable for the public.

The second method of obscuration does not involve disguising or hiding the means of egress as much as not showing the building occupants where the exits are. Illumination and marking of the means of egress with lighted exit signs is necessary if the public is to be expected to evacuate. With the exception of dwelling units and occupancies with small occupant loads and a single exit, most buildings and spaces will be equipped with internally illuminated exit lights and means of egress lighting connected to an emergency power source. The emergency power source may be a battery system or generator, but must pick up the assigned load within 10 seconds of power failure.

Emergency lighting is designed to enable the occupants to safely exit the building in the case of an emergency. The required intensity of 1 footcandle is not sufficient for much more than that. The 1–1½ hour duration required by the building codes is not designed to give the occupants time to finish shopping or eating before they evacuate the building.

COMPROMISES OF THE INTEGRITY OF CONSTRUCTION ELEMENTS

Many elements of the means of egress have specific fire-resistance rating requirements or operational requirements. Exit access corridors that serve thirty or more people, exits, smokeproof enclosures, and horizontal exits all have specific requirements for the construction of structural assemblies and opening protectives. The integrity of these elements can be compromised by tradesmen who penetrate rated assemblies while performing utility work, or by a tenant who innocently props open the stairway door to make it easier to reenter with his hands full. Both scenarios are common occurrences and both can have the same effect of making some part of the means of egress less effective or perhaps ineffective.

If a fire occurs on the same floor where our friend has propped open the stairway door and there are only two stairwells, we could conceivably lose 50 percent of the exit capacity of our building. What if the building is a high-rise? These are all "what-ifs," but if what-ifs did not occur, there would be no need for building or fire codes.

NOTE
Tenants are often unaware that their "security locking system" is subject to the building and fire codes.

Operational requirements include those for door hardware such as locking and latching, direction of door swing, and marking of interior stairway doors. Special locking arrangements and access controlled egress doors pose special problems in large office buildings. Tenants are often unaware that their "security locking system" is subject to the building and fire codes, and some security companies are either ignorant of the fact or choose to violate the code by installing locking systems outside of the construction permit process.

The installation of any device on a rated door assembly poses several problems. The obvious problem is that the function of the lock or device conflicts with the intent of the code (see Figure 11-4). All means of egress doors must be *openable from the side from which egress is to be made without the use of a key or special knowledge or effort.* That means one motion, on one knob, handle, paddle, or crash bar. Adding another locking arrangement may require occupants to perform two operations such as push on the paddle and then turn the knob. This is *not* acceptable.

NOTE
In addition to the operational problems that add-on devices may pose, they may violate the listing of the rated door assembly.

In addition to the operational problems add-on devices may pose, they may violate the listing of the rated door assembly. Unless the door and add-on device are listed for installation together, the add-on cannot be installed. If it has already been installed, the owner has real problems. NFPA 80 *Standard for Fire Doors and Windows* provides guidance for the installation of accessories on fire doors, as well as for their repair.

Operational requirements for special locking arrangements and access controlled egress doors can be found in a bit different form in each of the model building codes. For each system there is a list of specific requirements that must be met

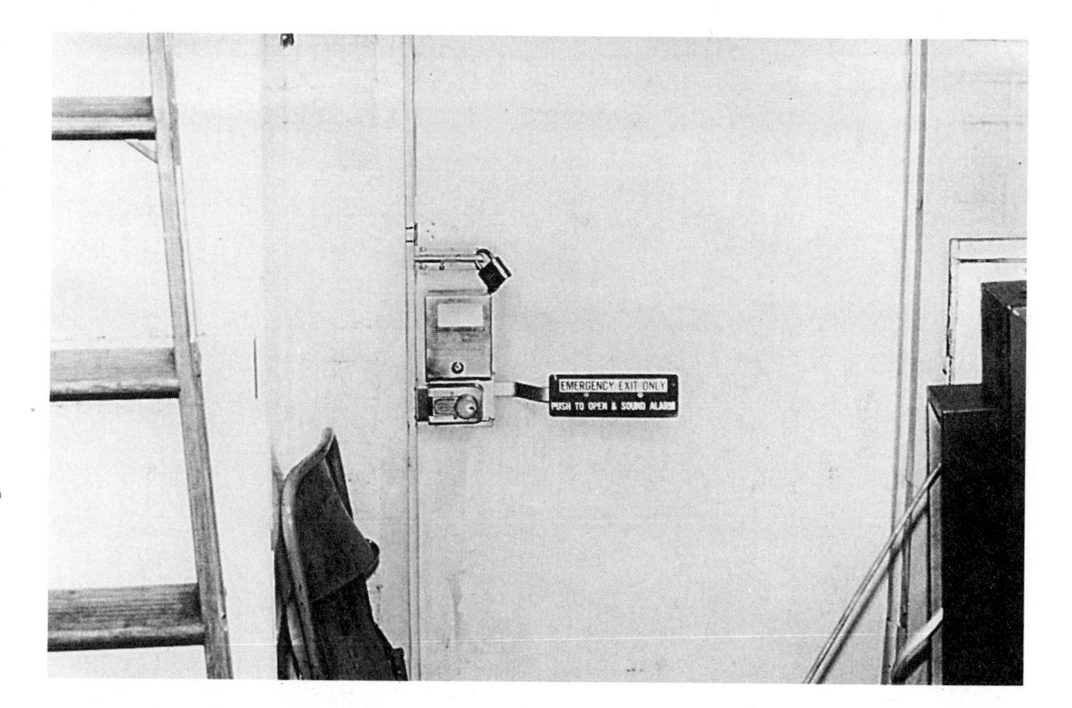

Figure 11-4 *Locks and security devices can create hazardous conditions. (Courtesy of Duane Perry.)*

in order to make use of the special provisions. In each case *all of the provisions must be met.* Use of the provisions is *not* a pick-and-choose exercise.

COMPROMISES OF THE INTEGRITY OF THE MEANS OF EGRESS FUNCTION

If posed the question "What are the functions of the building areas that comprise the three different parts of the means of egress?" most of us would answer in the same way, "To get people out of the building." We would be right, at least partly. The function of the exit access, exit, and exit discharge together is to get the people out and away from the building. The problem is that only one part of the means of egress, the *exit,* actually has a part of the building dedicated to it and absolutely nothing else (see Figure 11-5).

The *exit access* is the part of the building that leads to the exit and to where you work, play, eat, pray, shop, or do anything else people do in buildings. Exit access is merely a path through the usable part of the building. The width of this path varies depending on the occupant load, but this is clearly the most dangerous part of the means of egress. If there is a fire in the space, occupants will be exposed while in the exit access.

Exit discharge is the path from the termination of the exit to the public way, and often the part that building owners have the least control over. It is no less important than the other two and is more likely to be obstructed than used for other purposes. Like the exit access, the exit discharge is merely a dedicated path

Figure 11-5 *No storage of any kind is permitted in an exit.*

■ NOTE
Trash containers, delivered goods, and the belongings of evicted residents are only a few of the enemies of the exit discharge.

through a general area, not a dedicated building element like the exit. Trash containers, delivered goods, and the belongings of evicted residents are only a few of the enemies of the exit discharge.

The *exit* is the only one of the three parts with a specific portion of the building dedicated solely for its use. It may be of considerable size as in the case of a high-rise or covered mall building, or may be only the thickness of a door frame. When exits are used for any purpose other than the movement of people, a dangerous condition has been created. Storage that is combustible or that could become an obstruction can quickly make the exit unusable by the occupants.

■ NOTE
When exits are used for any purpose other than the movement of people, a dangerous condition has been created.

REMOVAL OR ELIMINATION OF EXITS

If the building permit process was used every time an addition was built or a temporary structure such as a tent was erected, there would be very few eliminated exits. Unfortunately, some people persist in trying to circumvent the process and what some would say is worse, it is usually the fire inspector who catches it—at the worst possible time.

Sometimes exits are eliminated when additions are added. What was once an exterior door in a restaurant now leads to a lovely glass-enclosed dining room with a balcony and no exterior stairs. If the work has been performed without a con-

■ NOTE
Buildings undergoing renovation during partial occupancy pose a particular challenge to fire and building inspectors.

■ NOTE
Occupants cannot be permitted to pass through construction areas while exiting the building; any work within the means of egress during business hours should be strictly prohibited.

■ NOTE
Exit discharges are especially threatened during construction due to the constant influx of construction materials, equipment, and trailers.

struction permit, there could be serious problems. Since an exit has been removed and the floor area has increased, exit capacity for the new occupant load is certainly questionable. Exit remoteness and travel distance could also have been adversely affected, not to mention the structural integrity of the balcony.

Another problem arises when occupants are forced to exit through an illegally constructed area that may not be structurally sound, or when legal construction has encroached upon an exit path making it a hazardous area. Assembly occupancies are famous for enclosing outside dining areas for the winter months. If the provisions of the building code and the construction permit process are followed, a safe, serviceable addition to the main structure may be possible. Unfortunately, prefabricated temporary structures are sometimes installed by contractors from outside the jurisdiction who have little regard for the code, permit process, or the manufacturer's installation instructions.

When these temporary structures are encountered by the fire inspector and the determination is made that the structural integrity has not been assessed by the building department, the fire inspector *should immediately issue an order that the addition be vacated.* Any exits from the existing structure that empty into the addition should be closed off and the building occupant load adjusted accordingly, until the building inspector gives the okay to reopen the new addition. The possibility of injury to the occupants from structural collapse in an uninspected temporary structure is real and necessitates immediate action.

Buildings undergoing renovation during partial occupancy pose a particular challenge to fire and building inspectors and require a close working relationship between the two departments. Provisions to ensure the adequacy and integrity of the required exits should be carefully outlined in preconstruction meetings for each phase of the project. Occupants cannot be permitted to pass through construction areas while exiting the building. *Any* work within the means of egress during business hours should be *strictly prohibited.*

Exit discharges are especially threatened during construction due to the constant influx of construction materials, equipment, and trailers. Buildings undergoing repair or renovation with partial occupancy require frequent spot checks on a regular basis so that fire prevention and occupant safety remain a priority of the general contractor.

SPECIAL REQUIREMENTS FOR BUILDINGS AND OCCUPANCIES

■ NOTE
Assembly occupancies have historically suffered the biggest loss of life due to problems with the means of egress.

Assembly Occupancies

Assembly occupancies have historically suffered the biggest loss of life due to problems with the means of egress. Many of our code requirements can be traced directly back to catastrophic fires in which a high loss of life led to a public demand for action. Many of the large fires we used in our discussions of interior finish and fire-resistive construction also had serious deficiencies in means of egress. The Coconut Grove and Beverly Hills Supper Club are two; the Happy Land Social

Club in New York City, where eighty-seven people were killed in 1990, was a more recent example.

Deficiencies in the means of egress were not the only problems in these buildings. They did ensure however, that the occupants were exposed to the deadly effects of fire for an excessive and, in many cases, fatal period of time. Because of our history in assembly occupancies, specific provisions for them have been included in the fire prevention codes.

NOTE
Overcrowding is not a subjective term; the codes define it as exceeding the occupant load calculated in accordance with the building code.

Posted Occupant Loads and Seating and Egress Plans

All public assemblies for fifty or more persons must have the approved occupant load posted. This is a key item to be checked during routine inspections. The worst time for an inspector to have to try to determine an occupant load is during a performance or during prime time at restaurants or nightclubs.

overcrowding
a condition in which the number of occupants within a building or space exceeds the approved occupant load calculated in accordance with the building code

Overcrowding is *not* a subjective term. The codes define overcrowding as exceeding the occupant load calculated in accordance with the building code. The fire prevention codes require the occupant load be posted and enforced. There should be no need for discussion between the business owner and inspector as to whether an unsafe condition exists. The code requires specific action by the fire official to ensure that the overcrowding condition is remedied *by the business owner*. It is the owner or manager who stops the show and takes the necessary steps to come within the posted occupant load. The inspector merely makes him do it.

NOTE
Overcrowding conditions, especially involving establishments that serve alcohol, can be particularly difficult and even dangerous for fire officials; personal safety must be a primary consideration.

Overcrowding conditions, especially involving establishments that serve alcohol, can be particularly difficult and even dangerous for fire officials. Personal safety must be a primary consideration. If police assistance is needed it should be requested, and if necessary the inspector should leave the premises until police assistance arrives. Meeting the business owner in his or her office, where there are fewer distractions has the advantage of allowing the owner to save face and thus be more cooperative, and also keeps potentially belligerent customers out of the picture.

NOTE
Many spaces have multiple occupant loads posted depending on whether the occupants are standing, seated at tables, or seated in rows of chairs.

As we discussed in Chapter 7, many spaces have multiple occupant loads posted depending on whether the occupants are standing, seated at tables, or seated in rows of chairs. Having multiple approved seating plans is an excellent way for public assemblies to have maximum flexibility in the use of their space and yet comply with the code.

Emergency Preparedness

The model fire codes address the issue of emergency preparedness by mandating the development of emergency plans, requiring personnel training in emergency evacuation and proper use of fire extinguishers, and requiring drills in high-risk occupancies. These facilities have large occupant loads, house hazardous processes, or contain persons whose capacity for self-preservation is reduced due to age, medical condition, or confinement. The fire codes also require accurate record keeping on the part of facility managers.

NFPA 1 and *NFPA 101* contain requirements for appropriately trained "crowd managers" at large events. The *IFC* contains requirements for fire evacuation plans, fire safety plans, emergency evacuation drills, and employee training and response procedures. *These provisions ensure prompt, appropriate actions on the part of building occupants in any emergency situation necessitating evacuation.* Evacuation is only part of the equation. Accurate accountability for employees and occupants after evacuation is also critical.

■ NOTE
Fire drills should be unannounced and should include scenarios that test the fire plan.

An evaluation of emergency plans, drills, and training records should be a part of each routine inspection. Plans and records must be routinely checked, or they will not be maintained by business owners.

Educational Occupancies

■ NOTE
Assembly areas in which children gather after evacuating a school must not be in areas through which responding fire equipment must pass.

Fire drills, considered a nuisance by some educators, are an important part of every schools' fire safety plan. Many fire safety professionals do not give school fire drills the attention that they deserve. Drills should be unannounced and should include scenarios that test the fire plan. An administrator might be posted at a particular corridor to inform teachers that for the purposes of the drill, the corridor is full of smoke and impassable. Administrators should also select a few students to remain in the office to see if they are reported as missing.

The time spent witnessing fire drills is productive time for fire inspectors and fire department operations units. The students, teachers, administrators, and parents realize that safety at the school is high on your list of priorities. Face-to-face meetings between operational units and school officials are also the best way to ensure adequate fire department access. Assembly areas in which the children gather after evacuating the school must not be in areas through which responding fire equipment must pass.

■ NOTE
Emergency plans and evacuation procedures for institutional occupancies must be based on the needs of the occupants and the design of the facility.

Records of all drills are required by the model fire prevention codes. As with all records required by the code, a review by the inspector during each routine inspection is mandatory.

Institutional Occupancies

■ NOTE
The emergency plan prepared by a facility can be an unintentional notification to the fire official of an illegal change in use.

Emergency plans and evacuation procedures for institutional occupancies must be based on the needs of the occupants and the design of the facility. Hospital or nursing home patients who are nonambulatory, or not capable of self-preservation, have very different needs than those in a halfway house. Both may be institutional occupancies, but the safeguards required by the building codes based on the different institutional use groups are quite different.

Every institutional emergency plan must address two key elements. The plan must adequately address the needs and limitations of the occupants, and the

■ NOTE
To safely use the "defend in place" strategy in institutional occupancies, they must be constructed with all of the safeguards required by the building code.

plan must be compatible with the construction features of the building. The emergency plan prepared by a facility can be an unintentional notification to the fire official of an illegal change in use. Occupancies that house persons who are incapable of self-preservation have much stricter construction requirements including fire-resistive construction, smoke partitions, and fire detection and fire sprinkler systems. To safely use the "defend in place" strategy in institutional occupancies, they must be constructed with all of the safeguards required by the building code. Approval of an emergency plan in which nonambulatory residents are instructed to remain in a building without the safety features prescribed by the model building codes could lead to disaster.

Participation in drills by fire department operations units is an invaluable test of the compatibility of the fire department preplan and emergency plan of the facility, especially in jails and prisons where firefighter safety takes on an added dimension.

■ NOTE
Participation in drills by fire department operations units is an invaluable test of the compatibility of the fire department preplan and emergency plan of the facility.

High-Rise Buildings

Fire safety and evacuation plans are required by all the fire prevention codes for high-rise buildings. Like all emergency plans, they should be "exercised" with the fire department. Fire drills are not required by the model codes for high-rise buildings, but they are a good idea. Advance planning and notification of all the building occupants by building management are required because most high-rise fire alarm systems selectively alarm specified floors and areas. Failure to make the necessary notifications can lead to confusion or fear for which the fire department will be blamed, regardless of fault.

Summary

The means of egress is a clear and unobstructed path that leads from any point within a building or structure to a dedicated public way or open space. It is comprised of three distinct elements called the exit access, the exit, and the exit discharge. Exit access and the exit discharge are merely paths, passages, or corridors through the otherwise occupied portions of the building. The exit however, is a space solely dedicated to the movement of the occupants. It may be expansive as in the case of exit stairways in a high-rise building or be no larger than the thickness of the door frame on a small building.

The exit access and exit discharge must be maintained in a clear and open condition with the appropriate width and height as determined by the building code. The exit must remain completely free of any storage or materials and is dedicated solely to the movement of the building occupants. There are five general ways in which the means of egress may be compromised: Exits are obstructed or, obscured, the integrity of their construction is compromised, the integrity of their function is compromised, or the exit is eliminated.

Review Questions

1. List the three elements of a means of egress.
   1. ______
   2. ______
   3. ______
2. List five ways in which the means of egress can be compromised.
   1. ______
   2. ______
   3. ______
   4. ______
   5. ______
3. List two ways in which the means of egress can be obstructed.
   1. ______
   2. ______
4. List two ways in which the means of egress can be obscured:
   1. ______
   2. ______
5. List two ways in which the integrity of construction elements and special features of the means of egress can be compromised.
   1. ______
   2. ______
6. List two ways in which the integrity of the means of egress function can be compromised.
   1. ______
   2. ______
7. List two instances in which exits may be eliminated or removed from service due to safety concerns.
   1. ______
   2. ______

8. Which occupancy classification has historically suffered the largest life loss due to problems with the means of egress?

____

9. When an overcrowding condition is encountered by the fire inspector, who is responsible for taking the necessary steps to comply with the approved occupant load?

____

10. All the model fire prevention codes require that employees of public assemblies receive training in _______________.

Discussion Question

1. You have been directed by the chief's office to look into a complaint by a city council member that a new house of worship is dangerously overcrowded on Friday afternoons. More than 300 cars have been counted in the undersized parking lot and within the surrounding community. The facility is an existing building formerly used as a school and is located within a residential neighborhood. Upon research you discover that the occupant load for the building based on square footage and capacity of the means of egress should be approximately five hundred persons. However, a zoning restriction imposed on the occupancy permit limits the facility to two hundred persons, based on the availability of only one hundred parking spaces in the parking lot.
   a. If on inspection, the facility in fact does have 500 persons in attendance, is it overcrowded?
   b. What occupant load should the fire official enforce?
   c. What would be your report to the chief's office?

Hazardous Materials

Learning Objectives

Upon completion of this chapter, you should be able to:

- Describe the control area concept of hazardous materials management.
- List the steps that must be taken to establish control areas.
- Describe the importance of hazardous materials management plans and hazardous materials inventory statements to the inspections process.
- Describe the NFPA 704 *Hazard Identification System.*

CODE PROVISIONS

The 1990s brought new code provisions for the storage, handling, and use of hazardous materials. The four model building and fire prevention codes were updated to better regulate hazardous materials. Some from the old school of fire prevention thought that the code process was being taken over by hazmat team members. Some even talked of leaving the enforcement of the *new* code provisions to the fire department hazmat team and only enforcing the *real* fire prevention regulations.

■ **Note**
For the purposes of the model building and fire prevention codes, the term hazardous materials is a laundry list of substances with potential physical and health hazards.

The facts speak a bit differently. The hazardous materials regulated by the reformatted and expanded codes are the same hazardous materials that the model fire prevention codes have regulated for years. Hazardous materials are not only things with the words *methyl-*, *ethyl-*, or *trichloro-*, at the beginning of their names. For the purposes of the model building and fire prevention codes, the term *hazardous materials* is a laundry list of substances with potential physical and health hazards. The list runs from combustible liquids, fibers, and dusts to oxidizers and water reactive materials; and from corrosives to toxic and highly toxic solids, liquids, and gases. For the most part, they are the same substances that we have regulated (with limited effectiveness) for many years.

The chapter or article in your fire prevention code entitled Hazardous Materials provides general regulations that apply to all hazardous materials as defined by the codes. Material-specific requirements are found within chapters or articles that address specific classes of substances such as flammable and combustible liquids or oxidizers.

■ **Note**
NFPA 704 establishes a marking system for buildings and tanks that provides immediate information as to the hazards to health, the flammability, the reactivity (instability), and special hazards of materials.

THE NFPA 704 SYSTEM

NFPA 704, *Identification of the Fire Hazards of Materials*, establishes a marking system for buildings and tanks that provides immediate information as to the hazards to health, the flammability, the reactivity (instability), and special hazards of materials (see Figures 12-1 and 12-2).[1] The *IFC*, *BNFPC*, and *SFPC* reference NFPA 704. The *UFC* adopts it by transcription as UFC Standard 79-3. The system uses a diamond divided into four color-coded quadrants. The numbers 0–4 indicate the relative hazards with the number 4 denoting the greatest hazard. The system is also used as a basis for the definition of *hazardous production materials* (HPM), which is discussed later in this chapter under the section Hazardous Production Material Facilities.

exempt amount
threshold quantity of a hazardous material established by the building and fire prevention codes as the maximum amount that can be stored, handled, or used within a building that is not classified as Use Group H

HAZARDOUS OR HIGH HAZARD USE GROUP BUILDINGS

In Chapter 3 we learned that Use Group H buildings are those in which quantities of hazardous materials exceeding the exempt amount/maximum allowable quantity (see Table 12-1)are stored, handled, or used. The construction requirements for H use buildings are extensive. Building height and area restrictions, separation requirements, fire protection, ventilation, and other special features are all expen-

Figure 12-1 *The NFPA 704 marking system warns responding units of potential hazards.*

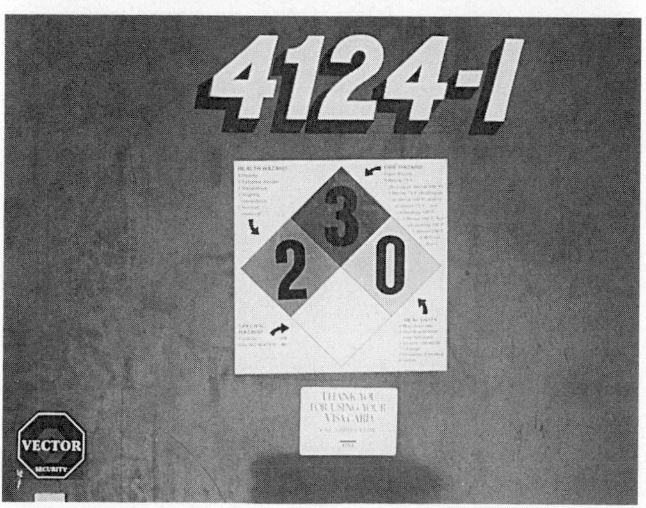

Figure 12-2 *Close-up of the NFPA 704 marking system.*

Table 12-1 *Maximum allowable quantity per control area [a] of hazardous materials posing a physical hazard*

Material	Class	Group When the Maximum Allowable Quantity Is Exceeded	Storage[b]			Use-Closed Systems[b]			Use-Open Systems[b]	
			Solid Pounds (cubic feet)	Liquid Gallons (pounds)	gas (cubic feet)	Solid Pounds (cubic feet)	Liquid Gallons (pounds)	Gas (cubic feet)	Solid Pounds (cubic feet)	Liquid Gallons (pounds)
Combustible liquid[c,i]	II	H-2 or H-3	N/A	120[d,e]	N/A	N/A	120[d]	N/A	N/A	30[d]
	IIIA	H-2 or H-3		330[d,e]			330[d]			80[d]
	IIIB			13,200[e,f]			13,200[e,f]			3,300[f]
Combustible fiber	Loose		(100)			(100)			(20)	
	Baled	H-3	(1,000)	N/A	N/A	(1,000)	N/A	N/A	(200)	N/A
Consumer fireworks (Class C, common)	1.4G	H-3	125[d,e,l]	N/A	N/A	N/A	N/A	N/A	N/A	N/A
Cryogenics, flammable		H-2	N/A	45[d]	N/A	N/A	45[d]	N/A	N/A	10[d]
Cryogenics, oxidizing	N/A	H-3	N/A	45[d]	N/A	N/A	45[d]	N/A	N/A	10[d]
Explosives		H-1	1[e,g]	(1)[e,g]	N/A	1/4[g]	(1/4)[g]	N/A	1/4[g]	(1/4)[g]
Flammable Gas	Gaseous			N/A	1,000[d,e]		N/A	1,000[d,e]		
	Liquefied	H-2	N/A	30[d,e]	N/A	N/A	30[d,e]	N/A	N/A	N/A
Flammable liquid[c]	1A			30[d,e]			30[d]			10[d]
	1B	H-2	N/A	60[d,e]	N/A	N/A	60[d]	N/A	N/A	15[d]
	1C	or H-3		90[d,e]			90[d]			20[d]
Combination (1A, 1B, 1C)		H-2 or H-3	N/A	120[d,e,h]	N/A	N/A	120[d,h]	N/A	N/A	30[d,h]
Flammable Solid		H-3	125[d,e]	N/A	N/A	125[d]	N/A	N/A	25[d]	N/A
Organic Peroxide	U[d]	H-1	1[e,f]	(1)[e,g]	N/A	1/4[g]	(1/4)[g]	N/A	1/4[g]	(1/4)[g]
	I	H-2	5[d,e]	(5)[d,e]	N/A	1[d]	(1)[d]	N/A	1[d]	(1)[d]
	II	H-3	50[d,e]	(50)[d,e]	N/A	50[d]	(50)[d]	N/A	10[d]	(10)[d]
	III	H-3	125[d,e]	(125)[d,e]	N/A	125[d]	(125)[d]	N/A	25[d]	(25)[d]
	IV		NL	NL	N/A	NL	NL	N/A	NL	NL
	V		NL	NL	N/A	NL	NL	N/A	NL	NL
Oxidizer	4	H-1	1[g]	(1)[e,g]	N/A	1/4[g]	(1/4)[g]	N/A	1/4[g]	(1/4)[g]
	3[k]	H-2	10[d,e]	(10)[d,e]	N/A	2[d]	(2)[d]	N/A	2[d]	(2)[d]
	2	H-3	250[d,e]	(250)[d,e]	N/A	250[d]	(250)[d]	N/A	50[d]	(50)[d]
	1	H-3	4,000[d,e]	(4,000)[d,e]	N/A	4,000[d]	(4,000)[d]	N/A	1,000[d]	(1,000)[d]

Table 12-1 *(Continued)*

Material	Class	Group When the Maximum Allowable Quantity Is Exceeded	Storage[b]			Use-Closed Systems[b]			Use-Open Systems[b]	
			Solid Pounds (cubic feet)	Liquid Gallons (pounds)	gas (cubic feet)	Solid Pounds (cubic feet)	Liquid Gallons (pounds)	Gas (cubic feet)	Solid Pounds (cubic feet)	Liquid Gallons (pounds)
Oxidizing gas	Gaseous		N/A	N/A	1,500[d,e]	N/A	N/A	1,500[d,e]	N/A	N/A
	Liquefied	H-3	N/A	15[d,e]	N/A	N/A	15[d,e]	N/A	N/A	N/A
Pyrophoric material		H-2	4[e,g]	(4)[e,g]	50[e,g]	1[g]	(1)[g]	10[e,g]	0	0
Unstable (reactive)	4	H-1	1[e,g]	(1)[e,g]	10[d,g]	1/4[g]	(1/4)[g]	2[e,g]	0.25[g]	(1/4)[g]
	3	H-1 or	5[d,e]	(5)[d,e]	50[d,e]	1[d]	(1)[d]	10[d,e]	1[d]	(1)[d]
	2	H-2	50[d,e]	(50)[d,e]	250[d,e]	50[d]	(50)[d]	250[d,e]	10[d]	(10)[d]
	1	H-3	NL	NL	NL	NL	NL	NL	NL	NL
Water reactive	3	H-2	5[d,e]	(5)[d,e]	N/A	5[d]	(5)[d]	N/A	1[d]	(1)[d]
	2	H-3	50[d,e]	(50)[d,e]	N/A	50[d]	(50)[d]	N/A	10[d]	(10)[d]
	1		NL	NL	N/A	NL	NL	N/A	NL	NL

Notes: For SI: 1 cubic foot = 0.023 m^3, 1 pound = 0.454 kg, 1 gallon = 3.785L.

NL = Not Limited; N/A = Not Applicable

a For use of control areas, see Section 414.2 [of the source document].

b The aggregate quantity in utilization and storage shall not exceed the quantity listed for storage.

c The quantities of alcoholic beverages in retail and wholesale sales occupancies shall not be limited provided the liquids are packaged in individual containers not exceeding 1.3 gallons. In retail and wholesale sales occupancies the quantities of medicines, foodstuffs, consumer or industrial products, and cosmetics containing not more than 50 percent by volume of water-miscible liquids with the remainder of the solutions not being flammable shall not be limited, provided that such materials are packaged in individual containers not exceeding 1.3 gallons (5L).

d Maximum quantities shall be increased 100 percent in buildings equipped throughout with an automatic sprinkler system in accordance with Section 903.3.1.1 [of the source document]. Where Note e also applies, the increase for both notes shall be applied accumulatively.

e Quantities shall be increased 100 percent when stored in approved cabinets, gas cabinets, exhausted enclosures, or safety cans as specified in the *International Fire Code*. Where Note d also applies, the increase for both notes shall be applied accumulatively.

f The permitted quantities shall not be limited in a building equipped throughout with an automatic sprinkler system in accordance with Section 903.3.1.1 [of the source document].

g Permitted only in buildings equipped throughout with an automatic sprinkler system in accordance with Section 903.3.1.1 [of the source document].

h Containing not more than the maximum allowable quantity per control area of Class I-A, Class I-B or Class I-C flammable liquids.

i Inside a building, the maximum capacity of a combustible liquid storage system that is connected to a fuel-oil piping system shall be 660 gallons provided such system conforms to the *International Fire Code*.

j Quantities in parenthesis indicate quantity units in parenthesis at the head of each column.

k A maximum quantity of 200 pounds of solid or 20 gallons of liquid Class 3 oxidizers is allowed when such materials are necessary for maintenance purposes, operation or sanitation of equipment. Storage containers and the manner of storage shall be approved.

l Net weight of the pyrotechnic composition of the fireworks. Where the net weight of the pyrotechnic composition of the fireworks is not known, 25 percent of the gross weight of the fireworks including packaging shall be used.

Source: 2000 International Building Code®, Table 307.7(1), pages 34–35. *Copyright 2000, International Code Council, Inc., Falls Church, Virginia. 2000 International Building Code. Reprinted with permission of the author. All rights reserved.*

Table 12-2 *Building code H Use Groups.*

Hazard	IBC	BNBN	SBC	UBC
Detonation	H-1	H-1	H-1	H-1
Deflagration	H-2	H-2	H-2	H-2
Physical hazard	H-3	H-3	H-3	H-3
Health hazard	H-4	H-4	H-4	H-7
Repair garages	S-1[a]	S-1[a]	S-1[a]	H-4
Aircraft hangars	S-1[a]	S-1[a]	S-1[a]	H-5
HPM facilities	H-5	F-1[a]	F-1[a]	H-6

[a] Special Use and Occupancy requirements apply.

sive propositions. The conversion of an existing Storage or Factory–Industrial Use building to an H is often impossible regardless of the financial resources available.

The model building and fire prevention codes have the exempt amounts of hazardous materials listed within tables, in slightly different formats. These quantities are used by design professionals at the front end of the construction process to properly classify a building according to the industrial processes that are to take place within the structure. Exempt quantities are used to limit the amount of regulated substances in buildings and structures that are *not* constructed with all the safeguards of a Use Group H building.

■ Note

Exempt quantities are used to limit the amount of regulated substances in buildings and structures that are not constructed with all the safeguards of a Use Group H building.

The moderate hazard storage facility is limited by the code to the exempt quantities of hazardous materials. To exceed those quantities, the building would have to undergo a change in use process and become a Use Group H building or comply with one or more of the exemptions that we shall discuss.

The three model building codes have slightly different provisions with regards to the H Use Group classification system. The *BNBC* and *SBC* have four H subgroups, the *IBC* has five, and the *UBC* has seven. As shown in Table 12-2, the model codes use slightly different systems to address the same issues. The *BNBC* and *SBC* use the basic provisions of use groups F and S for repair garages, aircraft hangars, and HPM facilities. These are then supplemented with Special Occupancy Requirements, much the same way high-rise buildings and covered malls are handled. The *IFC* and *UBC* established a separate use group for the facilities simplifying the process somewhat.

EXCEPTIONS TO THE H USE GROUP CLASSIFICATION

Inspections of H occupancies may require the inspector to bone up on the code a bit. Most jurisdictions just do not have that many H occupancies. What inspectors do commonly encounter, however, are buildings in which an *illegal change in use* has technically occurred. The moderate hazard factory and industrial occupancy in which quantities of hazardous materials exceeding the exempt amounts are stored or used presents a serious violation of the fire prevention codes.

The problem could be any of the substances regulated by the building and fire prevention code, and it might be several of them. The inspector is likely to hear "It has been this way for years," or "You will put me out of business." Both could well be true.

The model building and fire prevention codes have developed a series of provisions that work together to regulate hazardous materials in buildings other than H Use Groups aimed at providing an acceptable level of safety for the public and for fire department personnel who may be called to the building in the event of an emergency.

Table 12-1, from the *International Fire Code*, lists the exempt quantities of substances that pose physical hazards such as detonation, deflagration, and rapid combustion. All five model building codes use tables to specify exempt quantities. Exceptions to the H Use Groups are listed within the code text in the *IBC*, *BNBC*, *SBC*, and *NFPA 5000*, and as footnotes to the exempt quantity table in the *UBC*.

The interpretation of the word *exception* is important, and each of the codes is slightly different. Before you waste a lot of time digging through the H Use Group provisions, look at the exception first. Generally, the following occupancies and processes involving the storage, handling, or use of hazardous materials will not cause a building or structure to be classified as a Use Group H building, *even if the exempt quantities are exceeded*:

- Retail storage of alcoholic beverages in containers not exceeding 1 gallon.
- Storage of foodstuffs and cosmetics where the quantity of water-miscible flammable liquid is less than 50 percent in containers not exceeding 1 gallon.
- Mercantile storage in accordance with the fire prevention codes.
- Control areas maintained in accordance with the fire prevention codes.

This is not, by any means, a complete listing.

CONTROL AREAS

control areas
area within a building in which hazardous materials in quantities not exceeding the exempt amounts may be stored, handled, or used

The establishment of **control areas** is used by all the model building codes to manage the storage, use, or handling of hazardous materials. The concept recognizes the establishment of area separations as a method of controlling the hazards presented by the various hazardous materials. Control areas are often a viable alternative to an expensive change in use for the factory–industrial facility that needs specific quantities of hazardous materials in order to operate.

The concept of High Hazard or Hazardous uses based on threshold quantities of specific hazardous substances has been around for many years. The fun-

■ Note
Control areas are often a viable alternative to an expensive change in use.

■ Note
The control area concept allows the building owner to establish multiple control areas within a building and to store and use or handle up to the exempt amount within each one.

damental flaw in the system was that it applied to all buildings and structures, regardless of the size or construction type. The operator of a factory that covered an entire city block and was constructed of masonry, was limited to the exempt amounts of hazardous materials. The 2,500-square-foot wood frame building on the outskirts of town was limited to the exact same amounts.

The control area concept allows the building owner to establish multiple control areas within a building and to store and use or handle up to the exempt amount within each one. Although each of the model codes has slightly different provisions, the fundamentals are the same:

- The number of control areas permitted per building and per floor are specified within the building and fire prevention codes.
- The number permitted in mercantile occupancies is limited to two.
- Control areas must be separated from one another by fire separation assemblies with a minimum 1 hour fire-resistance rating.
- Ratings for floors and ceilings are also required.

In Figure 12-3, four control areas for the storage of Class II combustible liquid have been established within a factory–industrial building. In all three model building codes, the exempt amount of Class II liquids that can be stored in a non-

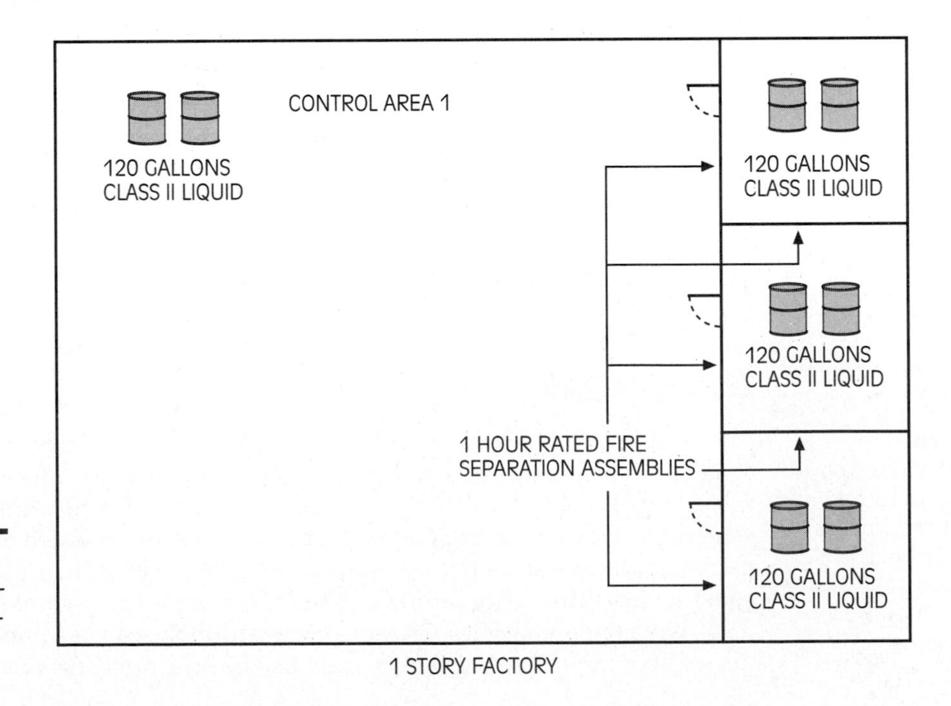

Figure 12-3 *Each control area is separated by fire separation assemblies rated for 1 hour.*

■ Note The establishment of control areas that involves any construction whatsoever requires the owner to secure a building permit.

■ Note The number of control areas is regulated per building, not per tenant.

■ Note A one-story building is limited to four control areas no matter how many tenants occupy the building.

sprinklered building is 120 gallons. Without the establishment of control areas, the factory would be limited to 120 gallons.

The codes would permit the establishment of four control areas in this one-story building, each containing up to the exempt amount. By separating the storage areas with fire separation assemblies, separate control areas are established and the factory can store up to 480 gallons of Class II combustible liquid.

Fire inspectors must remember that the establishment of control areas that involves any construction whatsoever requires the owner to secure a building permit. The fire inspector should direct the owner to the building department, so that construction plans can be reviewed, permits secured, and the installation can be inspected.

Another key item is that the number of control areas is regulated per *building*, not per *tenant*. Multiple tenants within single buildings are collectively limited to the same number of control areas as a single tenant who occupies the same space. A one-story building is limited to four control areas no matter how many tenants occupy the building. Determining who has property rights to the control areas is not a role nor is it a responsibility of the fire inspector. It is the responsibility of the building owner.

Control Areas in Mercantile Occupancies

Control areas within mercantile occupancies follow slightly different rules. Individual packages offered for retail sale are generally quite small. Most consumers do not pick up 55-gallon drums of corrosive materials, drop them into a shopping cart with the paper towels and mixed nuts, and stroll over to the cash register. In recognition of this fact, the codes permit mercantile occupancies to exceed the exempt amounts for storage per control area for certain materials, providing certain conditions are met.

These maximum quantities or "mercantile exempt amounts for storage and display," are significantly higher than the base exempt amounts. This exception applies only to the storage and display of *nonflammable solid* and *nonflammable* or *noncombustible liquids*. All other substances are limited to the normal exempt amounts. Additionally, individual containers and packages are limited in size, and displays are limited both in height and by the number of pounds or gallons per square foot. Increased aisle widths are also required.

■ Note In addition to the exempt amount for each class of hazardous material, the fire prevention codes have established a permit threshold, and the two are not interchangeable.

INSPECTIONS IN BUILDINGS WITH HAZARDOUS MATERIALS

A quick look at the building code exempt amount tables should bring about the realization that most buildings will contain at least some hazardous materials. The model fire prevention codes contain specific provisions aimed at the management of this inspections process. In addition to the exempt amount/maximum allowable quantity for each class of hazardous material, the fire prevention codes

Hazardous Materials Inventory Statement (HMIS)
an inventory of regulated materials in a form approved by the fire official that includes the following information: chemical name; manufacturer's name; hazardous ingredients; UN, NA, or CIS identification number; and maximum quantities stored or used on-site at any time

Hazardous Materials Management Plan (HMMP)
a management and contingency plan in a form approved by the fire official that includes the following information: site plan, floor plan, material compatibility, monitoring and security methods, hazard identification, inspection procedures, employee training, and emergency equipment available

Material Safety Data Sheet (MSDS)
a document, prepared in accordance with DOL 29 CFR, that contains information regarding the physical and health hazards associated with a given product or substance and a recommended emergency action

have established a permit threshold. The two are *not* interchangeable. The permit threshold establishes a level at which inspections for compliance with the code is considered prudent by the collective bodies of the model code organizations. There is no need to justify your inspection. As we discussed in Chapter 2, the model code process has justified it for you.

By using the permit system, you prioritize your inspections and simplify issues surrounding your right of entry. In dealing with hazardous materials code requirements, you also gain valuable information within the permit application process. Each of the model fire prevention codes permits the fire official to require the submission of a **Hazardous Materials Inventory Statement** (HMIS) and **Hazardous Material Management Plan** (HMMP). The *SFPC* and *UFC* include sample plans as appendixes. Copies of **Material Safety Data Sheets** (MSDS), which are submitted as a part of the HMIS, are also required to be available on site. Requirements for maintenance vary by code depending on the quantities of hazardous materials. Locking repository containers in approved locations are required in some occupancies.

A thorough and effective inspection of many occupancies is practically impossible without a review of the facility HMIS and HMMP. Imagine attempting to inspect a large warehouse-style hardware store. An inspector would be hard pressed to determine if the occupancy was within the exempt amounts for combustible liquids alone, much less for all of the hazardous materials. Think of all of the products that are classified as combustible liquids—paints, varnishes, cleaners, adhesives, waterproofing, pesticides, and many more. Consider that they are spread throughout the store on shelves and in the back storage areas. Now imagine going through this exercise for each class of regulated material! A review of the HMIS and perhaps a spot check of selected materials for accuracy of the HMIS, makes the goal of a thorough, complete inspection possible.

Fire officials should also consider requiring the use of pounds, gallons, and cubic feet for reporting quantities in HMIS. SARA Title III, Tier II reports list quantities in pounds for all substances regardless of the physical state. Since the model building and fire prevention codes only use weight for solid materials, inspectors are forced to spend precious time translating pounds to gallons for liquids, or pounds to cubic feet for compressed gases, depending on the substance.

Material Classification and Identification

A methodical, step-by-step approach must be used in determining what code requirements are applicable to a particular substance or product. Many products contain constituents that cause them to fall within several categories. Such materials must *comply with the provisions for each category.* Liquid pesticides that have a combustible liquid base must comply with the provisions for pesticides *and* combustible liquids. In determining conformance with the exempt amount/maximum allowable quantity restrictions, the pesticide would be included within the total quantity for toxic or highly toxic materials *and* the total quantity for combustible liquids.

■ Note
A thorough and effective inspection of many occupancies is practically impossible without a review of the facility HMIS and HMMP.

■ Note
Many products contain constituents that cause them to fall within several categories, and such materials must comply with the provisions for each category.

The first step in the process is a review of the MSDS. The NFPA 704 classification gives general guidance. Do not be surprised if the material poses three or more distinct hazards, each with its own exempt amount/maximum allowable quantity and permit threshold. Use the following process to determine the code requirements for a product:

1. Use the MSDS to determine the different hazards posed by the product (Class 2 Oxidizer, Flammable Liquid, Toxic Materials, etc.)
2. Using the fire prevention code, determine the permit threshold and the exempt quantity *for each hazard classification.*
3. Apply the general provisions of the article or chapter on hazardous materials.
4. Apply the material specific requirements from the applicable chapters or articles of the code.

Example

Chemical products are used in modern building ventilation systems to kill microorganisms that can grow within the system and be distributed throughout the building. One product in common use is classed as a Class 2 Oxidizer, Corrosive and Toxic Material, and is manufactured in the form of hockey-puck-size tablets. Table 12-3 lists the exempt quantities for the material based on the three hazard categories. Table 12-4 lists the permit thresholds for each.

Under the provisions of the *UFC*, a permit is required to store, use, or handle over 100 pounds of this product in any building or structure. Storage within a single control area is limited to 250 pounds. Keep in mind, however, that this example assumes that this is the *only* hazardous material within the building. To determine the need for permits and maximum quantity per control area, the total quantity of all materials that fall within a given category would have to be calculated. If a building owner stored fifty pounds of the product from our example and more than fifty pounds of another Class 2 Oxidizer, he would still need a permit. The *permit is per class of material*, not per specific product.

Storage and handling of the material must also comply with the provisions of all three classifications. Storage of our example product in the basement of a building is prohibited by all model codes under the provisions for Class 2 solid oxidizers. All the requirements for toxic materials and corrosives would also apply.

Table 12-3 *Exempt quantity (in pounds; nonsprinklered building).*

	IFC	BNFPC	SFPC	UFC
Class 2 Oxidizer	250	250	250	250
Corrosive	5,000	5,000	5,000	5,000
Toxic Material	500	500	500	500

Table 12-4 *Permit threshold (in pounds)*

	IFC	BNFPC	SFPC	UFC
Class 2 Oxidizer	100	50	100	100
Corrosive	1,000	1,000	1,000	500
Toxic Material	100	500	125	100

HAZARDOUS PRODUCTION MATERIAL FACILITIES

hazardous production material (HPM)
a solid, liquid, or gas that is classified as a 3 or 4 in accordance with NFPA 704 for hazards to health, flammability, or reactivity and used in research or production where the end product is not hazardous

Hazardous production materials have a health, flammability, or reactivity of Class 3 or 4 when rated in accordance with NFPA 704 or UFC 79-3. HPM facilities are research, laboratory, or production facilities that produce an end product that is not hazardous.

As indicated in Table 12-2, the *BNFPC* and *SFPC* classifies HPM facilities as factory–industrial buildings. The *IBC* and *UFC* establish a separate H Use Group. Semiconductor manufacturing plants are HPM facilities. Each of the model fire prevention codes has a specific chapter or article on HPM facilities. The materials storage and handling requirements are different than those for like materials in occupancies that are not HPM facilities. They quite literally have their own set of rules.

Summary

Code provisions for the storage, use, and handling of hazardous materials have been coordinated between the model building and fire prevention codes to provide alternatives to the High Hazard uses of the past. Materials that pose multiple hazards must comply with the code provisions for each hazard. Permit thresholds and exempt amount/maximum allowable quantity are applied on the basis of total quantity per category, not total quantity per product. The general provisions for all hazardous materials as well as the material specific requirements from throughout the code must all be applied.

Review Questions

1. Which standard establishes a marking system for buildings and tanks that provides information as to the health hazards, flammability, and reactivity of materials? ________
2. The maximum quantity of a regulated substance that can be stored within a control area in a building that is not classified as H or High Hazard is ________________.
3. Areas within a building in which hazardous materials in quantities not exceeding the exempt amount may be stored, handled, or used are called __________________.
4. The model building and fire prevention codes specify the maximum number of control areas per __________________.
5. List two documents that must be submitted with each hazardous material permit application.
   1. ________________________________
   2. ________________________________
6. A __________________ or document that describes the relative hazards must be on site for each hazardous material which is used, handled, or stored.
7. Are exempt quantities and permit or approval thresholds for hazardous materials the same? _____
8. List the three types of facilities that may be classified as HPM facilities.
   1. ________________________________
   2. ________________________________
   3. ________________________________
9. Materials that pose multiple hazards must comply with the requirements for

   __________________.
10. Permit or approval thresholds for hazardous materials that pose multiple hazards are based on __________________.

Discussion Questions

1. Based on a recommendation from the clerical staff, the new fire marshal is considering dropping the requirement that all permit applications for hazardous materials include

a Hazardous Materials Inventory Statement. Complaints of excess paperwork have been received from industry and the clerical staff.

a. How will this potentially affect the inspections process?

b. What is your recommendation to the fire marshal?

2. During an inspection of a nonsprinklered factory, you note that 750 pounds of a Class 2 oxidizing material are being stored in the building. The exempt amount is 250 pounds. The manager informs you that he cannot purchase the material from his supplier in quantities less than 500 pounds. The storage area is tidy and in good order and there are no other code violations in the building. What storage options are available to the factory manager under the provisions of the code?

Flammable Liquids and Aerosols

Learning Objectives

Upon completion of this chapter, you should be able to:

- Describe the system used to classify flammable and combustible liquids.
- Describe the hazards involved in the storage of aerosol products and the protection methods required by the model codes.
- Discuss the importance of using tanks in accordance with their listings.
- Discuss the importance of labeling requirements for containers.

flammable liquid
a liquid having a flash point below 100°F and further categorized based on flash point and boiling point as types IA, IB, and IC

combustible liquid
a liquid having a flash point at or above 100°F and further categorized based on flash point as types II, IIIA, and IIIB

NOTE
One in six fires that occurred in the United States between 1983 and 1987 involved flammable or combustible liquids, with losses exceeding 655 million dollars.

heat of combustion
the amount of heat given off by a particular substance during the combustion process; a measure of fuel efficiency

water-miscible
water soluble

flash point
the minimum temperature at which a liquid gives off sufficient vapor to form an ignitable mixture at the surface, but not sufficient to sustain combustion

Liquids that burn have been with us for a long time. Asphalt was used for roads and to waterproof boats in Mesopotamia prior to 3000 B.C. The ancient Egyptians used it in the mummification process to ensure that Amenhotep, Tutankhamen, and the other pharaohs passed into the afterlife in good condition. The Byzantines perfected the use of combustible liquid as a weapon in their defense of Constantinople in the seventh century A.D. *Greek fire*, a mixture of crude oil, charcoal, sulphur, and other chemicals was poured over the surface of the sea and ignited.

Today we drive our gas-guzzling cars on asphalt roads, heat our homes and businesses with fuel oil, and lubricate nearly all machinery with some type of combustible liquid. The average American drinks 2 gallons of beer and wine each year,[1] our food is prepared with combustible vegetable oils, and we spray or rub cosmetics containing alcohol on our bodies.

Flammable liquids and **combustible liquids** are very much a part of our everyday lives. They are also a major factor in our national fire experience. One in six fires that occurred in the United States between 1983 and 1987 involved flammable or combustible liquids. The losses exceeded 655 million dollars.[2]

PHYSICAL PROPERTIES

What is it about liquids that burn that makes them so dangerous? The very same properties that have made them so useful to us. Regardless of the source, most share certain characteristics. They are excellent fuels, the **heat of combustion** of kerosene is more than double that of wood.[3] They are plentiful, occurring naturally, as in petroleum, or through industrial processes or fermentation. Some, such as ethyl alcohol are **water-miscible**, whereas others are not and can be used as water repellents and sealants. They give off flammable vapors at comparatively low temperatures. The **flash point** of gasoline for instance is about –45°F.[4] Flammable and combustible liquids generally ignite at relatively low temperatures, making them readily available sources of heat energy, which is used to power engines, heat boilers and furnaces, and generate electricity. Imagine life without cars, cocktails, or hair spray! Not to mention heat and artificial lighting.

Many code requirements are based on the physical properties of the liquids. Liquids with flash points below normal ambient temperatures are naturally a greater hazard than those that must be heated in order for them to give off flammable vapors. A classification system based on physical characteristics is used by the model codes. Examples are shown in Table 13-1. Requirements for storage, handling, and transportation of flammable and combustible liquids are based on these classifications.

CODE PROVISIONS

Requirements for the storage, handling, and use of flammable and combustible liquids are found within the model fire prevention codes and referenced standards.

Table 13-1 *Liquid classifications.*

Classification	Flash Point	Boiling Point	Examples[a]
Class IA Flammable	< 73°F	< 100°F	Ethyl ether
Class IB Flammable	< 73°F	≥ 100°F	Ethyl alcohol gasoline
Class IC Flammable	≥ 73°F and < 100°F	N/A	Butyl ether
Class II Combustible	≥ 100°F	N/A	Fuel oil #1 (kerosene)
Class IIIA Combustible	≥ 140°F and < 200°F	N/A	Fuel oil #6
Class IIIB Combustible	≥ 200°F		Lubricating oil (motor oil)

[a] James H. Meidl, *Flammable Hazardous Materials* (Beverly Hills, CA: Glencoe Press, 1970), p. 44.

■ NOTE
Flammable and combustible liquids generally ignite at relatively low temperatures, making them readily available sources of heat energy.

NFPA 30, *Flammable and Combustible Liquids Code*, NFPA 30A, *Automotive and Marine Service Station Code*, and NFPA 30B, *Manufacture and Storage of Aerosol Products*, are referenced by the *BNFPC* and the *SFPC*. The *UFC* contains all applicable provisions within the code and applicable *UFC* Standards. Several other standards relating to tank vehicles, oil burning equipment, and repair garages are also referenced by the different model codes. Information for the installation and maintenance of oil burning equipment is also found in the model mechanical codes.

GENERAL FIRE SAFETY REQUIREMENTS FOR LIQUIDS

The misuse and improper handling of flammable and combustible liquids have resulted in staggering losses of lives and property. A fire incident involving a gasoline tanker that overturns on the interstate gets a fair amount of play in the press. The hapless citizen who suffers severe burns from trying to clean paint brushes with gasoline usually does not even make the local paper. You can bet that the paint brush cleaning incidents vastly outnumber the tanker crashes. The model fire prevention codes contain general fire safety requirements in an attempt to protect the public from their own actions and those of others. General safety requirements in all the model codes include:

- All containers (see Figure 13-1), vessels, and tanks containing flammable or combustible liquids must be approved for the purpose and marked or labeled with an appropriate warning.

Figure 13-1
Approved containers must be labeled in accordance with DOT requirements.

- *All* equipment, piping, and fittings must be listed and approved for the specific application, including dispensing apparatus, pumps, and valves.
- Dispensing into unapproved containers is prohibited.
- The sale of Class I liquids as domestic cleaning solvents is expressly prohibited. Class I and II liquids may not be used indoors for parts cleaning except within machinery listed for that purpose.
- Flammable and combustible liquids must be dispensed in an approved manner. Flammable liquids must be drawn from tanks by pump. Combustible liquids may be dispensed by gravity through approved equipment.
- Flammable and combustible liquids may not be discharged onto the ground or into any waterway. All waste must be disposed of in an approved manner.

container
a vessel with a capacity of 60 gallons or less used for the storage or transportation of flammable or combustible liquids (piping and engine fuel tanks are not containers)

STORAGE

Containers and Portable Tanks

Storage requirements are based on the type of vessel and classification of the liquid. Code provisions for containers and portable tanks address **containers** up to

■ NOTE
All containers and portable tanks must be tested and listed for the intended service.

60 gallon and portable tanks with up to 660 gallon capacity (see Table 13-2). All containers and portable tanks must be tested and listed for the intended service. Standards for the design of metal drums, safety cans, and plastic containers have been developed by the Department of Transportation (DOT), the American National Standards Institute (ANSI), the American Society for Testing and Materials (ASTM), and Underwriters Laboratories (UL). The appropriate standards are referenced in your model fire prevention code.

Quantity limitations within occupancies are based on the use group of the building, class liquid, and storage arrangements including type of sprinkler pro-

Table 13-2 *Container and portable tank sizes are limited by NFPA 30.*

Liquids Container Type	Flammable Liquids Class IA	Combustible Class IB	Class IC	Class II	Class III
Glass	1 pt	1 qt	1 gal	1 gal	5 gal
Metal (other than DOT drums) or approved plastic	1 gal	5 gal	5 gal	5 gal	5 gal
Safety cans	2 gal	5 gal	5 gal	5 gal	5 gal
Metal drum (DOT Specification)	60 gal	60 gal	60 gal	60 gal	60 gal
Approved metal portable tanks	660 gal	660 gal	660 gal	660 gal	660 gal
Polyethylenes DOT Spec. 34, U.N. 1H1, or as authorized by DOT Exemption	1 gal	5 gal	5 gal	60 gal	60 gal
Fiber drum NMFC or UFC Type 2A, Types 3A, 3B-H, or 3B-L, or Type 4A	—	—	—	60 gal	60 gal

Note: SI Units: 1 pt = 0.473 L; 1 qt = 0.95 L; 1 gal = 3.8 L.

Source: Reprinted with permission from NFPA 30, *Flammable and Combustible Liquids Code*, copyright © 2000, National Fire Protection Association, Quincy, MA 02269. This reprinted material is not the complete and official position of the National Fire Protection Association on the referenced subject which is represented only by the standard in its entirety.

■ NOTE
Special sprinkler requirements apply to the storage of flammable and combustible liquids, separate from those found within NFPA 13 and UBC 9-1.

tection in the building. As we discussed in Chapter 6, special sprinkler requirements apply to the storage of flammable and combustible liquids, separate from those found within NFPA 13 and UBC 9-1. These requirements are incorporated by reference by the *BNFPC* and *SFPC* from NFPA 30, *Flammable and Combustible Liquids Code*. Article 79 of the *UFC* includes the requirements within the code text.

The storage of Class I liquids in basements is expressly prohibited. The storage of Class II liquids in basements is prohibited by the *SBC* unless the area is protected with a sprinkler system complying with the requirements of NFPA 30, and such storage is severely restricted by the *BNFPC* and *UFC*.

Aboveground Storage Tanks

AST
aboveground storage tank for regulated liquids

UST
underground storage tank for regulated liquids

NFPA 30 and *UFC* Article 79 include siting requirements for aboveground tanks. The distance tanks can be located from property lines depends on the size of the tank and exposure protection. All tanks, whether aboveground (**AST**) or underground (**UST**), must conform to one of the standards referenced by the model fire prevention codes shown in Table 13-3.

As an inspector, you will encounter tanks with capacities that range from 275 gallons to several million gallons. Do not think that *bigger* necessarily means more challenging in gaining code compliance. The manager of the tank farm probably understands that code compliance is a necessary part of doing business. Mr. Wizard, the local garage owner, who has invented his own revolutionary system of doing oil changes by dispensing motor oil with air pressure from a cannibalized home heating oil tank, is probably a bigger challenge. Mr. Wizard has violated the requirement of using tanks and equipment in accordance with their listings. The home heating oil tank is the lightest service tank made. He has pressurized a vessel that was not designed to be pressurized, risking catastrophic rupture. Add to the rupture possibility the fact that pressurized air, with 21 percent oxygen, will be forcing a combustible liquid out of the ruptured tank. Hopefully, you will find Mr. Wizard's invention before your friends in the engine company have to respond to his garage for the fire.

Table 13-3 *Tank design standards.*

Tank Construction	Standard
Fiberglass reinforced plastic and polyester	UL 1316, ASTM D4021
Steel	API 12B, API 12D, API 12F, API 650, UL 58, UL 80, UL 132, UL 443, STI Standard for Dual Wall Steel USTs

■ **NOTE**
Any vessel with a capacity of 60 gallons or less is a container.

Types of Aboveground Storage Tanks ASTs range in size from just over 60 gallons to several million gallons. By definition, any vessel with a capacity of 60 gallons or less is a container. Large tanks at storage facilities include floating roof tanks that may be open or covered, cone roof tanks, dome roof tanks, and horizontal tanks.

ASTs at most other facilities include unprotected tanks up to 660 gallons used for the storage of fuel oil connected to the building services, portable tanks of up to 660 gallons (see Figure 13-2), tanks installed in special enclosures in accordance with NFPA 30A, and concrete-encased fire-resistant tanks (see Figure 13-3).

Inspection of Aboveground Storage Tanks Whether you are inspecting a tank farm with a dozen internal floating roof tanks that have an aggregate capacity of 20 million gallons or a 6,000-gallon concrete-encased fireresistant AST, your inspection involves checking for compliance with the same basic requirements. Small ASTs of ≤ 6,000 gallon capacity are becoming a popular replacement for USTs for the storage of motor fuels and fuel oil for building services. Many building owners are unaware that permits and inspections are required for their installation. Some contractors are also unaware of the requirements or choose to ignore them. Inspectors should pay particular attention when encountering new ASTs. *Was the installation permitted and approved?* A secondary consideration must also be,

Figure 13-2 *This unprotected steel tank lacks impact protection. Placed on unlevel ground, its secondary containment is inadequate.*

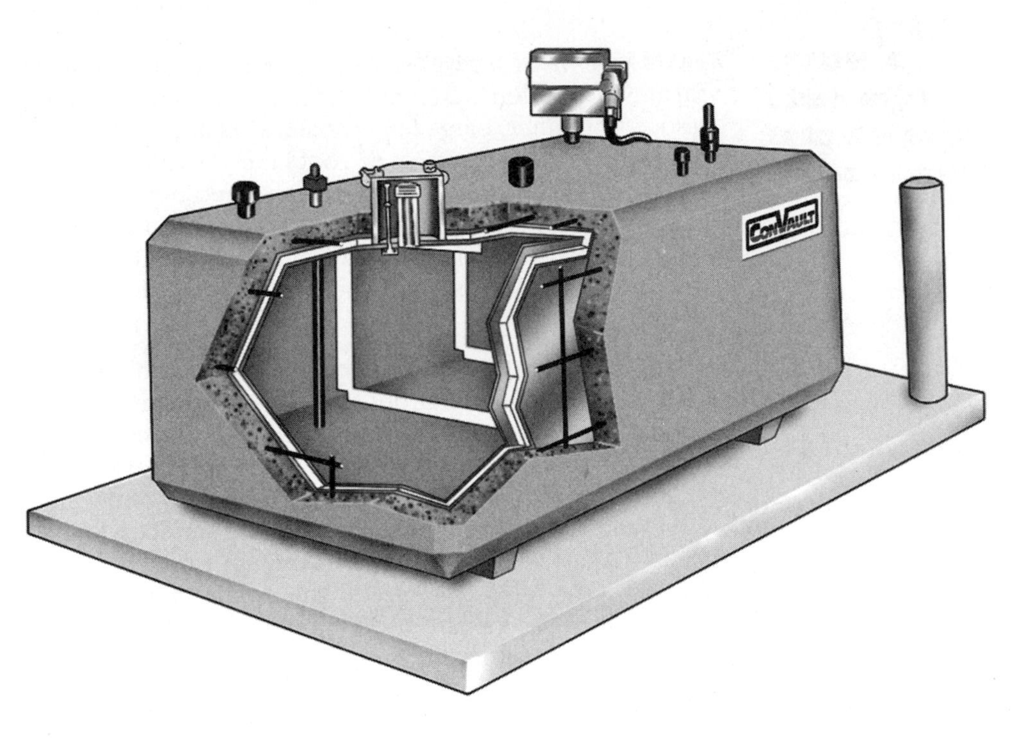

Figure 13-3 *This concrete encased tank incorporates fire resistance, leak detection, and secondary containment and is projectile resistant. (Courtesy of ConVault Mid-Atlantic, Inc.)*

what happened to the tank that was replaced. *Was it abandoned underground also without a permit and in violation of the code?*

■ **NOTE**
Physical protection for ASTs and piping from vehicular damage is required.

- *Hazard identification* is required for stationary ASTs using the NFPA 704 system. Signs do not have to be mounted directly on tanks, but must be readily visible from outside of diked areas. Physical protection for ASTs and piping from vehicular damage is required. The *BNFPC* and *SFPC* set a performance standard for bollards or barriers. They must be able to resist a force of 12,000 pounds applied 30 inches above the ground surface. The *IFC* and *UFC* prescribe guard posts of not less than 4-inch diameter steel filled with concrete, buried to a depth of at least 3 feet and extending 3 feet above the ground surface.
- *Supports, anchorage, and foundations* that support tanks must be stable and not subject to undermining from environmental forces nor subject to vehicular damage.

■ **NOTE**
Any changes to the diking system should be designed by a licensed engineer and reviewed by the fire official.

- *Dikes* around tanks must be maintained free of vegetation and in good repair. Dikes are designed to impound liquid spills due to overfill or tank failure and are designed to contain a spill equal to the capacity of the largest tank within the diked area. Dikes are an engineered fire protection feature. Any changes to the diking system should be designed by a licensed engineer and reviewed by the fire official.

- *Vents* must be clear and unobstructed and must not terminate in an area that would pose a hazard. Construction that occurs after the installation and approval of an AST can create problems when construction encroaches on vents.
- *Control of spillage* and housekeeping are key items that protect against damage to the environment and accidental ignition.
- *Records for the required maintenance and testing* of fixed fire protection systems, fire detection and alarm systems, and fire protection equipment should be checked to verify compliance with the testing requirements of the applicable referenced standards.
- *Inventory records* should be checked to verify that inventory control practices are in place and are followed by facility personnel. As with all records required by the model fire prevention codes, they must be checked by fire officials during each routine inspection. Failure to ensure that records are complete and in an acceptable form will lead to lax or no record keeping at the facility.

Underground Storage Tanks

USTs must be installed in accordance with the requirements of the model codes, applicable referenced standards, and the manufacturer's installation specifications. USTs are tested at the factory and should be tested with air pressure at 3–5 psi prior to being placed within the excavation. Once the tank is in place and secured to any required tie downs, a third air pressure test must be performed in the presence of the fire official, prior to backfilling. These tests ensure that the tank was not damaged in transport or damaged by the contractor during installation. The three tests are complementary and none should be waived by the fire official. They provide assurance to the public and to the owner of the facility that the UST was tight at installation. Piping must also be tested with air pressure for tightness in accordance with your model fire prevention code.

NOTE
Under no circumstances should an air test be performed on a tank that contains a flammable or combustible liquid.

On a rare occasion, an inspector will encounter a tank that was installed and filled without the required tightness tests or inspections. Under no circumstances should an air test be performed on a tank that contains a flammable or combustible liquid, or that has contained a flammable or combustible liquid. Imagine several thousand cubic feet of compressed air forcing flammable vapors out of a small opening in a tank. Think of the static electrical charge building as the fuel–air mixture rushes out of the tank. Imagine the contractor's explanation after the accident. "I just did what the fire inspector ordered."

Procedures for testing tanks that have contained flammable and combustible liquids are contained within standards referenced by the model codes and should be followed to ensure safety. The contractor should not be released from the requirement that a tightness test be performed. He should be required to submit for approval an alternate test procedure that meets the intent of the tightness test and complies with good engineering practice. In this way, the owner and the public have assurance that the tank is tight and the contractor learns his lesson by having to enlist the service of a consulting engineer or testing contractor.

closure
tank closure; placing a tank permanently out of service by removal or abandonment in place using an approved method

Routine UST Inspection Routine UST inspection involves a review of inventory control records and procedures, a visual inspection of exposed vent and vapor recovery piping, and assessment of general fire prevention and spill control practices. Tanks that are out of service must be closed by removal or abandonment in place. Tank **closure** procedures should be strictly observed to guard against underground petroleum releases. Petroleum in the groundwater is a significant fire, health, and environmental hazard.

■ **NOTE**
Tank closure procedures should be strictly observed to guard against underground petroleum releases.

SERVICE STATIONS AND GARAGES

In addition to compliance with code requirements for tank storage, service stations and garages must comply with specific code requirements relating to the dispensing of motor fuels and repair of vehicles. As shown in Table 13-4, the model fire prevention codes differ in their approach to organizing the requirements, although the basic requirements are very similar.

Supervision is required at all service stations open to the public by the *BNFPC* and *SFPC*. Attendants must be capable of responding to emergencies at the site, operating emergency shutoff controls and fire protection equipment, and handling routine spills. The attendant is also responsible for preventing the illegal dispensing of Class I liquids into unapproved containers.

■ **NOTE**
The service station attendant is also responsible for preventing the illegal dispensing of Class I liquids into unapproved containers.

Dispensers

Dispensers, commonly referred to as "pumps," should be carefully inspected. Public exposure to the hazards associated with flammable and combustible liquids is probably the greatest at the dispenser island of their local service station.

■ **NOTE**
Public exposure to the hazards associated with flammable and combustible liquids is probably the greatest at the dispenser island of their local service station.

- *Dispenser protection* must be provided by installing barrier protection or raising the dispensers on a 6-inch island. Lots that have been repaved, raising the surface to within 6 inches of the top of dispenser islands must be restored to the original level or have barrier protection installed, or the islands must be raised.
- *Hoses, nozzles, and fittings* must be listed and in good condition. Removal of automatic hold-open levers by station operators in violation of the listing voids the listing on the nozzle.

Table 13-4 *Code requirements.*

	IFC	BNFPC	SFPC	UFC	NFPA 1
Flammable and combustible liquids	Chapter 34	Chapter 32	Chapter 9	Article 79	Chapter 28
Service stations	Chapter 22	Chapter 32	Chapter 9	Article 52	Chapter 22
Repair garages	Chapter 22	Chapter 32	Chapter 35	Article 29	Chapter 22

- *Emergency shutoff features* that control power to the pumps and the flow of fuel should be inspected. A remote master shutoff that controls electric power to the pumps must be accessible and prominently marked. An approved emergency shutoff valve, which automatically closes in the event of fire or impact, must be installed in each dispenser. Valves must be tested at the time of installation and be retested by the operator annually.

AEROSOL STORAGE

aerosol
a product that is dispensed from a container by means of a liquified or compressed gas

In 1982, a fire at the Kmart distribution center in Bucks County, Pennsylvania, demonstrated the potential hazards of the storage of **aerosol** products. The 1.2-million-square-foot facility was fully sprinklered. Storage was palletized and on racks and was 15 feet high. In-rack sprinklers were not provided.

A case of aerosol carburetor cleaner was knocked off a pallet and fell to the floor. An employee operating a propane powered forklift reported hearing a hissing sound following immediately by flames. He attempted to reach a fire extinguisher, but was driven back by the fire, which had extended up the face of the stack from which the case had originally fallen. Other employees attempted to extinguish the fire but were also forced to evacuate.

The fire, fueled by the 40 to 50 pallets of carburetor cleaner in the vicinity of the point of ignition broke through the roof within 13 minutes. Ruptured aerosol containers rocketed through the structure, trailing burning contents and rapidly spreading the fire. The fire burned out of control for 8½ hours and was not totally extinguished for 8 days. Property damage alone exceeded $100 million, and worse, Kmart's distribution system had been destroyed.[5]

Full-scale fire tests conducted during the late 1970s and early 1980s demonstrated the potential for rapid fire development associated with the rupture and rocketing of aerosol containers.[6] In many of the tests, conventional sprinkler protection was quickly overwhelmed. Standard storage arrangements including aisle widths and travel distances were inadequate. As a result, specific requirements for the storage of aerosols were incorporated in the model fire prevention and building codes and NFPA 30B, *Code for the Manufacture and Storage of Aerosol Products*, was developed.

Aerosol Classifications

■ NOTE
Each carton of aerosol product must be clearly marked on at least one side with the words "Level ___ Aerosols."

Aerosol products are classified Level 1, 2, or 3, based on the chemical heat of combustion of the base product and the propellant. Storage requirements are based on this classification. Each carton of aerosol product must be clearly marked on at least one side with the words "Level __ Aerosols." Unmarked cartons are automatically considered to be Level 3, or the most hazardous. Products such as whipped cream, in which neither the base product nor the propellant are flammable, are not classified and are not regulated under these code sections.

Storage Requirements

nonsegregated storage the storage of aerosol products in general purpose warehouses within areas that are not used exclusively for the storage of aerosols

segregated storage the storage of aerosol products within areas specifically designed for that use incorporating code-required storage arrangements and fire protection features

There are five basic storage configurations for aerosol products. Each has specific requirements for quantity limits, storage configuration, and fire protection.

1. *Aerosol or liquid warehouses* are special structures specifically used for the storage of flammable liquids and aerosol products. The *IFC* and *UFC* use the term *aerosol warehouse* and the *BNFPC* and *SFPC* refer to NFPA 30, which has requirements for *liquid warehouses.*

2. *General purpose warehouses* are buildings where various commodities are stored within a single building. Distribution centers for hardware and supermarket chains as well as other mercantile occupancies are general purpose warehouses. For the purposes of aerosol storage requirements, storage areas in mercantile buildings are also general purpose warehouses. Storage arrangements as well as quantities often depend on the season. Deck stain and patio furniture do not sell well in Minnesota in January.

Two storage arrangements are permitted in general purpose warehouses. **Nonsegregated storage**, in which aerosol products are stored throughout the facility intermixed with other commodities, is regulated depending on storage arrangement (palletized or rack storage). Quantities permitted within **segregated storage** areas are significantly higher. Segregated storage areas are enclosed by rated walls or chain-link fence enclosures and are protected with sprinkler systems designed to apply water at much higher densities than required by NFPA 13 or UBC 9-1.

■ NOTE Segregated storage areas are enclosed by rated walls or chain-link fence enclosures.

3. *Retail sales areas* are limited to specific quantities depending on the sprinkler protection provided. Permitted quantities are increased if the area is sprinklered. An additional increase is given if the sprinkler system is designed to protect areas used for the storage of aerosols.

4. *Storage in A, B, E, F, I, and R use buildings* is limited to 1,000 pounds of Level 2; 500 pounds of Level 3; or 1,000 pounds of Levels 2 and 3 combined.

5. *Outdoor storage* of aerosol products, including products stored in trailers or temporary buildings, must be located with respect to exposures, including buildings and lot lines, and exit discharges.

■ NOTE Low flash point liquids, especially when atomized during spray finishing operations, are particularly unforgiving.

APPLICATION OF FLAMMABLE FINISHES

■ NOTE Inspectors should question signs of overspray on garage floors and caution against spraying in unauthorized areas.

Flammable finishes are applied to vehicles, furniture, and various other consumer items through spraying or dipping operations. The result is a durable finish that is hard to match using nonflammable products. The use of low flash point liquids results in rapid drying and ease of multiple applications. Low flash point liquids, especially when atomized during spray finishing operations, are particularly unforgiving.

spray booth
a structure designed and constructed to be used for the application of flammable finishes, featuring power ventilation, fixed fire suppression, and separation that is installed within a building

spray room
a room designed and constructed for the application of flammable finishes complying with the requirements of the building code

spray area
an area designed and constructed to be used for the application of flammable finishes complying with the requirements of the building code

Spray Finishing

Spray finishing must be conducted in an area designed and constructed for that purpose. Spray finishing is specifically prohibited in A, E, I, or R occupancy buildings, unless within a properly constructed spray room. Spray finishing is only permitted within approved **spray booths**, **spray rooms**, or **spray areas**.

At auto body repair facilities employees sometimes spray outside spray booths during particularly busy times. This is strictly prohibited and extremely dangerous. Power sanding, grinding, and other spark-producing activities provide a ready ignition source. Inspectors should question signs of overspray on garage floors and caution against spraying in unauthorized areas.

Inspections of spray finish facilities should include attention to the following conditions: Paint and solvent storage must be in approved cabinets or in a room constructed for that purpose. Liquids must be in approved safety cans or approved containers. Quantities must comply with the code requirements. Approved self-closing waste cans (see Figure 13-4) must be provided. Flammable liquid containers must be bonded to prevent against static discharges. Fire doors that protect paint storage rooms are commonly propped open or the closer assemblies are removed. These rated opening protectives must be maintained in accordance with the code.

Approved fire extinguishers should be visible, accessible, and maintained in

Figure 13-4
Approved waste cans must be provided and used! (Courtesy of Ron Berry.)

■ NOTE
Particular attention should be paid to portable fire extinguishing equipment in auto body shops—it gets used!

■ NOTE
Sprinkler heads should be protected with thin plastic bags that are frequently changed.

Approved fire extinguishers should be visible, accessible, and maintained in accordance with NFPA 10 or UFC 10-1. Particular attention should be paid to portable fire extinguishing equipment in auto body shops—it gets used!

Spray booths must be separated from combustible storage or construction by a 3-foot clear area, or a rated wall or partition. Filters must be clean and permit the designed air flow through the ventilation system. Discarded filters must be immediately removed to a safe detached location or stored submerged in water until discarded. Booths should be kept clean and free of residue, and sprinkler heads should be protected with thin plastic bags that are frequently changed.

Spray rooms must be maintained in a clean condition. Combustible storage within the spray room is not permitted. Inspectors should ensure that filters are clean and that the ventilation system is operating efficiently. Residue buildup on ventilating fans quickly destroys their efficiency. All ventilation systems must discharge to a safe location, away from ignition sources, other air intakes, building openings, or exit discharges.

Dipping Operations

Dip tanks, into which items are submerged to apply flammable finishes, must be installed within areas that comply with building code requirements. Depending on the size of the tank, an overflow system, emergency bottom drain, automatic closing cover, and fixed fire suppression may be required. Open flames, spark-producing equipment, or unapproved electrical equipment are prohibited within the vapor area.

Inspectors should ensure that approved self-closing waste cans are provided and that waste material is disposed of properly. Suppression systems should be maintained in accordance with appropriate standards and their listings. Ventilation equipment must operate efficiently and discharge to safe locations.

Automobile Undercoating

The spray application of automobile undercoating is permitted to be conducted outside of approved spray booths or spray rooms only if adequate natural or mechanical ventilation is provided and the undercoating materials have a flash point $\geq 100°F$.

DRY CLEANING

Code requirements for the construction, operation, and maintenance of dry cleaning plants depend on the class solvent used in the dry cleaning process. Dry cleaning establishments range from pick-up and delivery sites, to self-service facilities that use nonflammable solvents, to dry cleaning plants that use combustible solvents. Building code provisions for use group classification and special occupancy requirements, as well as the special requirements of the fire prevention codes, are based on solvent classification as shown in the following list:

Class Solvent	*Flash Point*
I	< 100°F
II	≥ 100°F
IIIA	≥ 140°F and < 200°F
IIIB	≥ 200°F
IV	Nonflammable

NOTE
The model fire prevention codes prohibit the change of solvents to another classification without prior approval by the fire official.

The model fire prevention codes prohibit the change of solvents to another classification without prior approval by the fire official. Perhaps as important, the machinery in use must be listed for the solvent. All tanks, piping, and equipment must be listed for the intended use.

Summary

The same physical characteristics that make flammable and combustible liquids useful make them treacherous. They emit flammable vapors at low temperatures, often well below normal ambient temperature. They have high heat of combustion ratings, often double that of wood and common combustibles.

The hazards associated with flammable and combustible liquids are managed by the codes by regulating specific areas:

- All equipment, containers, tanks, and piping must be listed for the intended use. This minimizes fires and accidents resulting from equipment and container failure.
- Quantities of liquids are limited according to building use and fire protection features.
- Ignition potential is minimized by prohibiting open flames, spark-producing processes, and unapproved electrical equipment.

Review Questions

1. What two physical properties of flammable and combustible liquids are used to classify them?
   1. ______________________________
   2. ______________________________
2. The amount of heat given off by a particular substance during the combustion process is called ______________.
3. The minimum temperature at which a liquid gives off sufficient vapor to form an ignitable mixture at the surface but not sufficient to sustain combustion is called its ______________.
4. Name the standard developed by the National Fire Protection Association to regulate the manufacture, storage, and handling of aerosol products. ______________
5. ______________ liquids may be dispensed by gravity from approved equipment.
6. Special sprinkler requirements for flammable liquid storage facilities can be found in ______.
7. A container is any vessel with a capacity equal to or less than ______.
8. Stationary tanks must be identified using the ________ hazard identification system.
9. Underground storage tanks (USTs) are tested with air at ________ psi prior to being placed within the excavation.
10. The storage arrangement in which aerosol products are stored throughout the facility intermixed with other commodities is called ______________ storage.

Discussion Question

1. During an inspection of a service station you observe a customer dispensing one dollar's worth of gasoline into a plastic antifreeze jug. He explains that he has run out of gas and his car is two blocks away. The attendant told him that without a twenty dollar deposit he could not borrow the metal safety can the station maintains for just this reason. On questioning, the attendant verifies the customer's statements. The station has lost three safety cans in four weeks to customers who have run out of gas. Besides, the antifreeze jug is listed for this use since antifreeze is combustible.
   a. Should you permit the customer to proceed to his car with the gasoline?
   b. What violation(s) of the code have occurred?
   c. Is the attendant right about the antifreeze jug's listing?
   d. What would be your actions?

Detonation and Deflagration Hazards

Learning Objectives

Upon completion of this chapter, you should be able to:

- List and describe the three basic categories of explosive materials in commercial use.
- Describe two different methods for the initiation of explosive materials.
- Describe the duties and responsibilities of the fire official in allegations of property damage or accidents involving commercial blasting.
- Describe the duties and responsibilities of the fire official in permitting public fireworks displays.
- Describe the handling of misfires of explosive materials and fireworks.

Depending on whose book you read, the first explosive material was developed by the Chinese, Arabs, or Byzantines. A monk named Roger Bacon first wrote the formula for black powder in the thirteenth century. Another monk, Berthold Schwartz, put Bacon's formula to use and invented the gun a century later. He was remembered by the epitaph: "Here lies Berthold the Black, the most abominable of inhumans, who by his invention has brought misery to the rest of humanity." And so, black powder got its name not from the monk who discovered it, but from a monk named Berthold,[1] who developed a practical use for the discovery.

■ NOTE
Black powder has been described as "treacherous," because its ignition temperature is so low that the slightest spark will ignite it at normal temperatures.

By the seventeenth century, black powder was used in mining throughout Europe. The first recorded use of black powder in Colonial America was in 1773 when a copper mine in Connecticut was converted into the Newgate Prison. In 1802 Eleuthere duPont deNemours began the commercial production of black powder near Wilmington, Delaware. Eleuthere lost his life in an explosion at the mill but members of his family continued production at what eventually became the chemical manufacturing giant DuPont.

deflagration
very rapid but subsonic oxidation evolving heat, light, and a low-energy pressure wave that is capable of causing damage

Black powder has been described as "treacherous," because its ignition temperature is so low that the slightest spark will ignite it at normal temperatures.[2] Black powder **deflagrates**, or burns very rapidly but at subsonic speeds. Black powder opened up the Midwest by blasting through the Appalachian and Rocky mountains so that the railroads could cross. Fatal accidents were a continual reminder of the unpredictability of black powder. A man and a mule were said to have been lost to accidents every day in crossing the Appalachians. Most of the laborers were recent immigrants to the United States. The railroad deeply regretted the loss of the mules.

Nitroglycerine was discovered in 1846 by an Italian scientist named Ascanio Sobrero who was so appalled by its destructive power that he cautioned against its industrial use.[3] In 1862 Alfred Nobel took over his father's company, which manufactured a nitroglycerine headache remedy called Glonoin Oil.[4] He began testing nitroglycerine for use in blasting, and although his brother was killed during the testing, he refused to give up. By 1866 he had discovered a method of using diatomaceous earth to absorb the nitroglycerine, thus making it reasonably stable. He named his invention *dynamite* after the Greek *dynamis* or power.

■ NOTE
Unlike black powder, a static spark or even a burning fuse will not initiate dynamite.

In addition to the ease of handling and relative stability of Nobel's invention, a key safety factor was the means of initiation. Unlike black powder, a static spark or even a burning fuse will not initiate dynamite. It takes a small explosion to start the explosive reaction. Early **detonators** were gun (percussion) caps loaded with fulminate of mercury, which were fired by means of a lit fuse. Today detonators are manufactured with such precision that delay blasting, in which multiple charges are fired at the same time and initiate explosive reactions milliseconds apart is the norm.

detonator
a device, consisting of electric and nonelectric blasting caps, fuse caps, and detonating cord delay connectors, that contains a primary or initiating explosive designed to set off an explosive reaction

explosive materials
explosives, blasting agents, and detonators, including dynamites, slurries, emulsions and water gels, black powder, smokeless powder, detonators and safety fuses, squibs, detonating cord, and other materials whose primary function is to function by explosion

explosives
chemical compounds or mixtures whose primary function is to function by explosion and that cause a sudden and almost instantaneous release of pressure, gas, and heat

magazine
a structure designed and constructed for the storage of explosive materials

blasting agents
materials or mixtures containing fuel and an oxidizer not otherwise classified as an explosive and that cannot be detonated by means of a #8 blasting cap when unconfined

EXPLOSIVE MATERIALS

Explosive materials (see Figure 14-1) is a catchall term that includes explosives, blasting agents, black powder, detonators, or blasting caps and other products, some of which are simply formulations of the preceding list. The relative safety of these materials varies greatly, as do the requirements for storage and handling. **Explosives** storage is limited to **magazines** with strict requirements for construction. **Blasting agents** are commonly stored in the same 40-foot semitrailer that transported them to the site. The erection of a fence and a padlock for the trailer are often the only required precautions.

The role of the fire inspector in regulating the storage, use, and transportation of explosives is normally limited to the areas of construction blasting, demolition, and special industrial explosives use. Mining and quarrying operations, by far the largest users of commercial explosive materials, are generally regulated by state government mining authorities. Interstate and interjurisdictional transportation of explosive materials falls under the authority of the Department of Transportation and state police agencies.

Permits are required by the fire prevention codes for the manufacturing, use, sales, display, possession, disposal, or transportation of explosive materials. Some states and some jurisdictions have established blaster certification programs similar to those established by state mine regulators. Blasters are tested and issued li-

Figure 14-1
Explosive materials. (Courtesy of Duane Perry.)

■ **NOTE**
Blasting agents are commonly stored in the same 40-foot semitrailer that transported them to the site.

censes or certificates of competency. Where tests have been properly developed and are technically valid, such programs are often embraced by industry as well as the public as a means of ensuring competent blasters.

Unfortunately, competency testing generally ensures the *minimum* acceptable level of performance. Explosives are particularly unforgiving, and minimal competence is never enough. Effective regulation forces the industry to establish internal procedures to ensure compliance, as well as promoting increased technical competence. Lax enforcement or occasional enforcement merely ensure that when an accident does occur, the public and elected officials will rightly ask what provisions were in place to protect them and who dropped the ball. Effective regulation is impossible without a basic knowledge of the process.

■ **NOTE**
Unfortunately, competency testing generally ensures the minimum acceptable level of performance.

BLASTING THEORY

Commercial blasting operations (see Figure 14-2) use explosive materials to break up or shatter a material. Generally rock formations are in the way of the road cut, or water or sewer line, or perhaps the basement of a new home. Holes are drilled in the rock to a specified depth in a specific pattern. The drill pattern, as well as the type and amount of explosives are elements of the blast design that must be determined by the blaster. The blast energy from a given amount of explosive material might shatter a hard rock such as blue stone. The same amount of explosive

Figure 14-2 *Holes are drilled into subsurface rock formations and explosives are used to break the rock for excavation. (Courtesy of Howard Bailey.)*

NOTE Less explosive material does not always mean less potential damage.

energy might be absorbed by a softer material such as shale, and then transmitted on through the ground to cause property damage. Less explosive material does not always mean less potential damage.

Amount of Explosives per Detonation

NOTE Scaled distance formulas are the simplest method of controlling potential damage.

The amount of explosive material that is used within a single shot depends on the type and amount of rock or other material to be broken up, the depth of the formation, the amount of dirt or cover that rests on top of the rock formation, and proximity to other structures, roads, or utilities. As early as 1934, when E. H. Rockwell developed his Energy Formula,[5] studies on damage caused by blasting operations have resulted in various formulas that limit charge weight or amount of explosive material detonated per shot.

NFPA 495, *Explosive Materials Code*, includes provisions to limit potential damage by limiting the amount of explosives detonated at one time or by monitoring the effects of the blasting with scientific instruments and maintaining blast effects within a safe range.

scaled distance formula a formula used to determine maximum amount of explosive material that can be detonated per delay interval of 8 milliseconds or greater, based on distance to the nearest occupied structure

Scaled distance formulas are the simplest method of controlling potential damage. NFPA 495[6] provides formulas that determine amount of explosives permitted based on the distance to the nearest occupied building. When blasting within 300 feet of the nearest building, the formula W (lb) = (D (ft)/50)^2 is prescribed, where W is the weight of explosives in pounds and D is the distance to the nearest occupied structure in feet. The maximum quantities are per delay interval of 8 milliseconds or more. This means that if charges are set to detonate at least 8 milliseconds apart, they can be considered distinct detonations.

Initiation of the Charge

detonator, instantaneous type a detonator with no time lapse between the firing signal and detonation of the main explosive charge

detonator, delay type a detonator that introduces a specific time lapse between the firing signal and detonation of the main explosive charge

The method of initiation is critical and depends on the type of detonators used. Detonators, commonly called *blasting caps* are inserted into explosives cartridges. The detonators may be either instantaneous or delay types. **Instantaneous type detonators** initiate a single charge per detonator. They may be electric or nonelectric, but the entire load of explosive material is detonated at once.

Delay type detonators usually delay detonation between adjacent boreholes between 10 and 60 milliseconds. This separation of explosive reactions, however slight, enables the blaster to use the same charge weight dependent on distance as used in instantaneous blasting, except per delay instead of per shot. In the examples in Figure 14-3, a blaster has ten holes drilled and is 200 feet from the nearest building. Using the scaled distance formula W (lb)= (D (ft)/50)^2, he would determine his charge weight dependent on distance as follows:

$$\begin{aligned} W &= (200/50)^2 \\ &= 4^2 \\ &= 16 \end{aligned}$$

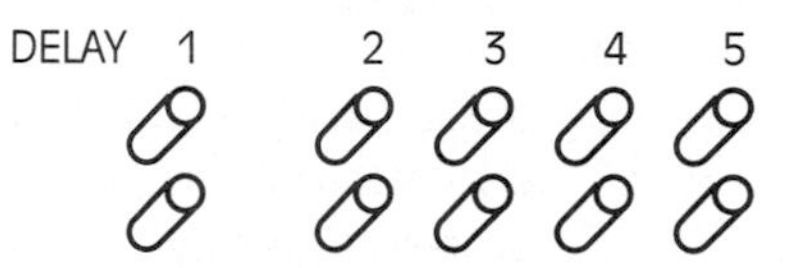

Figure 14-3 *By using delay type detonators, more explosives can safely be detonated per shot.*

■ NOTE

The benefits of delay blasting are obvious; blasters can get the job done with fewer shots and with less potential property damage.

The blaster would be limited to 16 pounds of explosive material per delay of 8 milliseconds or more, meaning 1.6 pounds per hole using instantaneous detonators where all ten holes would fire simultaneously. Using the same example except with delay type detonators, the blaster could use up to 16 pounds per delay, or 8 pounds per borehole because only two fire simultaneously. The benefits of delay blasting are obvious. Blasters can get the job done with fewer shots and with less potential property damage.

Blasting Agents

■ NOTE

Blasting agents are explosive materials that are not cap sensitive under normal conditions of storage and use.

primer
explosives packages made up of an explosive charge and a detonator or detonating cord used to initiate other less sensitive explosives or blasting agents

Blasting agents are explosive materials that are not cap sensitive under normal conditions of storage and use. Commonly known by the acronym ANFO (for ammonium nitrate + fuel oil), blasting agents are a relatively cheap, safe, and efficient method of blasting. Because blasting agents are not cap sensitive they must be used in conjunction with high explosives that are. A high explosive **primer** is used to initiate the blasting agent.

Blasters in the field commonly refer to explosives-grade prilled ammonium nitrate as "fertilizer," but do not kid yourself. You would not want to put this stuff on your garden. The use of ammonium nitrate in explosives was patented in Sweden in 1847.[7] One day in 1921, workers at Badische Anilin und Soda Fabrik plant in Oppau, Germany, did what they had done over 16,000 times before when they needed some ammonium nitrate. It was stored outside, exposed to the weather in a 4,500-ton crusted mound. They drilled holes and set explosives charges. Something that day was different. When the smoke cleared, a 50-foot deep crater 450 feet in diameter was all that was left of what was the factory. More than 400 people were dead and the shock wave was felt 145 miles away.[8]

■ NOTE
When determining charge weight per delay, one pound of blasting agent equals one pound of explosives.

When determining charge weight per delay, one pound of blasting agent equals one pound of explosives.

Blasting Safeguards

stemming
inert material placed in the borehole after the explosive material has been loaded to confine the effects of the explosive reaction; may also separate charges within a single borehole

flyrock
rock propelled from the blast area by blasting operations

The type and amount of cover and proper **stemming** are key elements in reducing the possibility of damage or injury due to **flyrock**. The first question that must be answered at any blasting operation is where will the rock break to? Rock that is hammered by explosives at supersonic speed and shattered moves. Without a free face or area of relief to break to, the rock will travel in the path of least resistance—up. This is not a big deal on a highway construction job a mile from any development. *It is a big deal* in populated areas or near utilities, railroads, or highways that are open.

■ NOTE
Without a free face or area of relief to break to, the rock will travel in the path of least resistance—up.

The amount of cover or dirt that covers a rock formation is extremely important. Without adequate earthen cover, rock cannot be safely blasted. Blasting mats (see Figure 14-4), which are used to help confine the effects of the blast, are not a substitute for adequate earthen cover. They are an enhancement, not a replacement.

Without adequate stemming, a borehole is a bit like a loaded cannon. You have a charge at the closed end and an explosive reaction with nowhere to go but out the barrel. Effective stemming confines the effects of the blast and maximizes rock breakage. It also minimizes flyrock. Rock drill cuttings from the boreholes makes poor stemming material. Clean, crushed stone is the most effective material. It must be carefully shoveled into the borehole after loading to avoid any damage to the electrical leg wires of the detonator (see Figure 14-5).

Figure 14-4 *Blasting mats prevent rock from being thrown from the blast area.*

Figure 14-5 *Care must be taken not to damage the detonator leg wires while stemming boreholes.*

■ **NOTE**
Effective stemming confines the effects of the blast and maximizes rock breakage.

How much stemming is also very important. The formula

$$T = .7 \text{ to } 1.3 \times B$$

where T is stemming height and B is the horizontal distance between boreholes is the textbook standard (See Figure 14-6).[9] A rough rule of thumb used in quarries

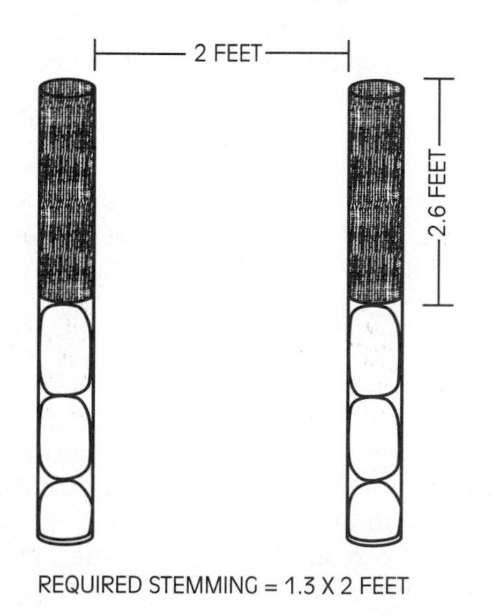

Figure 14-6 *The amount of stemming required is based on the distance between bore holes.*

and surface mines is *twice the diameter of the borehole in feet.* It is a little heavy with the small diameter holes used in construction blasting, but it gets you close.

CONSTRUCTION BLASTING

Shortly after three o'clock on the morning of November 29, 1988, six firefighters were killed in a series of explosions at a construction site. Two fires, one involving a pickup truck and one involving a semitrailer containing blasting agents, were found by arriving crews. While attempting to extinguish the fire in the semitrailer, it exploded, killing the crews of both engine companies. Approximately 40 minutes later, a second trailer exploded, fortunately without additional injuries.[10] Other incidents where firefighters have been killed fighting fires in semitrailers containing explosive materials occurred in 1959 and 1964. Nineteen people, including four firefighters were lost in the two incidents.[11]

Explosive materials are used in road and utility construction across the United States every day. The object is to break the rock into pieces that are easily dug by heavy equipment. Blasting activity depends on the amount and type of underground rock formations within the jurisdiction and the economic pressure to develop. The cost associated with the removal of underground rock may shift development to areas where blasting operations are not necessary. Unfortunately, this sometimes leads to spotty development.

■ NOTE **Blasting operations associated with development adjacent to established neighborhoods is one of the most challenging regulatory exercises a fire official will ever experience.**

Tracts of vacant, often wooded land that require blasting to put in every sewer line or dig every basement, end up surrounded by established communities. People in the established communities frankly are not particularly interested in any new development. The need for new housing ended abruptly the day their new home was finished. They do not want the construction traffic. They thought that land was park land, after all—and they do not want their house to be damaged by blasting operations. Blasting operations associated with development adjacent to established neighborhoods is one of the most challenging regulatory exercises a fire official will ever experience.

Permits for the storage, use, transportation, sale, disposal, and manufacture of explosives are required under the model fire prevention codes. Permits enable the fire official to keep track of all operations and activities involving explosive materials, and allow the fire official to establish administrative guidelines as *conditions of permit.*

■ NOTE **The storage of explosives materials in any quantity is restricted to approved magazines.**

Explosives Storage

The storage of explosive materials in any quantity is restricted to approved magazines. Magazine types and construction can be found in each of the model fire prevention codes, NFPA 495, and in the Institute for Makers of Explosives (IME) Safety Publications Library. There are five types of magazines (types 1–5), each with a particular use and limits as to the quantities and types of explosives that may be stored. Requirements range from extensive, for type 1 magazines (weather-, bullet-, and fire-resistant with redundant locking arrangements) to minimal (weather-resistant and lockable) for day boxes.

Magazine location and storage capacity are determined by compliance with the American Table of Distances approved by IME. The table sets maximum storage amounts in magazines depending on distance from other magazines, inhabited structures, and rail lines and highways. Quantities are increased (significantly at the low end of the table) when the magazine is effectively barricaded. Artificial barricades must be mounds or revetted walls of earth a minimum of 3 feet thick. Natural barricades are simply the natural features of the area and may be a combination of the terrain and timber such that surrounding exposures cannot be seen from the magazine when the trees are bare of leaves.

Storage within magazines is closely regulated by the fire prevention codes. Explosives and detonators must be stored in different magazines. Storage requirements for high explosives and blasting agents are different. Although blasting agents can be stored in portable trailer type magazines, high explosives cannot. Strict requirements for lighting, locking, and security are contained within the codes.

Problems associated with the storage of explosive materials at construction sites can be a challenge for the fire official. Problems with security and potential fire exposure are common. These conditions are easier to control at permanent storage facilities such as those maintained at mines and quarries. Blasting operations may be the responsibility of a subcontractor who does not control the movement and storage of large amounts of fuel and combustible materials that are stored and used in the operation. Regardless, the blasting contractor is responsible for the safe storage and handling of explosive materials on the site.

■ NOTE
The blasting contractor is responsible for the safe storage and handling of explosive materials on the site.

If the general contractor is storing combustible materials too close to the magazines, the blasting contractor only has two choices. Get the situation fixed or move the explosives to another site. Inspectors should remember that the role of the fire prevention code official is to ensure the safety of the public through the effective enforcement of the fire prevention code, not the arbitration of disputes between general contractors or subcontractors. Permits for the storage and use of explosive materials should be immediately revoked upon a finding of unsafe or reckless storage use or handling. Second chances are few with explosive materials.

■ NOTE
Permits for the storage and use of explosive materials should be immediately revoked upon finding of unsafe or reckless storage use or handling.

Explosives Transportation

The interjurisdictional transportation of explosives is generally the responsibility of the Department of Transportation and state police agencies. Transportation within individual jurisdictions is addressed by the model fire prevention codes and is serious business. Explosives and detonators can only be transported on the same truck when the procedures of DOT 49 CFR *Specification for the Transportation of Explosive and Other Dangerous Articles* are followed (see Figure 14-7). Detonators and explosives are segregated by hauling the detonators in a steel box lined with 1-inch thick oak.

Figure 14-7
Explosives are transported to construction sites by truck. (Courtesy of Howard Bailey.)

Trucks hauling explosives between the magazine and job sites must travel by the least congested routes. Trucks traveling in the same direction must maintain a 300-foot distance between each other. Trucks must be mechanically sound, equipped with fire extinguishers, and not carry unauthorized passengers.

Blasting Operations

Prior to the issuance of a permit for the storage or use of explosives, a thorough inspection of the site should be made. Permit applications should include a map or scale diagram of the site indicating the location of occupied structures, utilities, and highways or rail lines.

Operators of utilities such as water, gas, electricity, and telecommunications must be notified by the blasting contractor at least 24 hours in advance. The utility may decide to have a representative on site during the operation. If blasting is conducted within the utility company's right-of-way, they may have requirements that exceed those of the fire prevention code. In such cases the blaster must comply with the utility company's wishes, although *you* cannot enforce the utility company's regulations. You are there to ensure compliance with the fire prevention code.

■ NOTE
It is not unreasonable to require contractors to notify the occupants of homes that are in close proximity to the blast area.

It is not unreasonable to require contractors to notify the occupants of homes that are in close proximity to the blast area. Blasting complaints can often be minimized if neighboring homeowners know in advance how long the operation will be and are aware that there is government oversight of the blasting operation.

■ NOTE
Nationally recognized good practice dictates that the blaster should have records for each shot.

Blast Records Blast records are not specifically required by the model fire prevention codes, although they are specified in NFPA 495. Nationally recognized good practice dictates that the blaster should have records for each shot. The investigation of any accident or allegation of property damage as a result of blasting will depend on the quantity and quality of information recorded by the blaster. For every shot the following minimum information should be recorded:

- Contractor name, job location, date, and time of day
- Type of rock, number of boreholes, and spacing
- Diameter and depth of boreholes
- Type and amount of explosives
- Amount of explosives per delay of 8 milliseconds or greater
- Method of firing and type of circuit
- Direction and distance to nearest occupied structure
- Weather conditions
- Earthen cover and (or) mats used
- Type of detonators and delay periods
- Type and height of stemming
- Seismograph records (if used)

■ NOTE
Good records are the best (and only) defense against allegations of damage by surrounding property owners.

Good records are the best (and only) defense against allegations of damage by surrounding property owners. Most unfounded allegations of damage are not made by dishonest people seeking to take advantage of the blaster. People often become aware that there is a potential for blast damage and *then* notice that crack in the concrete basement floor for the first time.

Resolution of these allegations are the responsibility of trained investigators hired by the blaster's insurance carrier. Blast engineers with extensive training and education are able to determine if damage is the result of blasting. The fire inspector's responsibility is to ensure that operations are conducted in accordance with the code. The blaster is probably liable for the damage (depending on state liability laws), whether he was operating within the code or not.

Misfires and Accidents Misfires and accidents are unusual, but inspectors must be prepared to deal with them. Misfires can occur when the insulation on the leg wire of an electric detonator is skinned, or shifting rock "cuts-off" a hole in the middle

■ NOTE
Misfires must be handled in a methodical manner with extreme caution.

of a shot, causing only part of the loaded boreholes to fire. Misfires can also occur if the shot is wired incorrectly or if detonators simply malfunction. Misfires must be handled in a methodical manner with extreme caution:

- The area must be barricaded to ensure that no persons gain access.
- The power source used to fire the shot must be disconnected and electric wires shunted (twisted together) or made safe.
- All personnel should stay out of the area for at least 1 hour if fuse caps are used and ½ hour for other type detonators.
- Only the blaster should then reenter and inspect.
- If possible, the remaining holes should be reconnected and fired.
- If necessary, stemming should be removed by hand, a detonator should be inserted, and the hole fired from the top of the column down.

■ NOTE
A fire inspector should never enter the blast area after a misfire.

If these methods are ineffective, outside help may have to be called in. The role of the fire official in misfires is to see that they are handled safely and expeditiously and that a minimum number of persons are put at risk. A fire inspector should never enter the blast area after a misfire.

Accidents involving explosives generally are investigated by state or federal occupational safety and health officials and law enforcement agencies tasked with explosives investigations. In some jurisdictions it is the fire prevention bureau or fire marshal's office. The inspector should see that the scene is safe and secured to prevent the destruction of evidence. Like the fire scene, the scene of a blasting accident is often full of activity by emergency personnel. Rescue workers may treat and transport the injured, leaving a path of boot prints and medical supply wrappers. The inspector should make every attempt to prevent unnecessary destruction of the scene.

Other Explosive Materials

Small arms ammunition and reloading supplies such as smokeless powder, black sporting powder, and small arms primers are commonly sold to hobby shooters and sportsmen. Industrial explosives and explosives actuated power tools are also commonly found in construction and in industrial applications.

Retail sales and display of smokeless powder (gunpowder), commercially manufactured black sporting powder, and small arms primers are limited. If stored within indoor portable magazines, up to 400 pounds of smokeless powder and 25 pounds of black sporting powder may be permitted to be stored in a fully sprinklered building. Displays are limited to 25 pounds of smokeless powder in original containers. The display of black sporting powder is expressly prohibited.

Mercantile occupancies can and should be guided into compliance with the code. The challenge posed by gun and hobby shows is enormous. Many reloaders and black powder enthusiasts attend these exhibitions, and vendors are only too

NOTE
Without strict guidelines for the operators of exhibition facilities and regular inspections during the hours of show operations, exhibit hall operators will not attempt to police their tenants.

willing to supply them with materials. Without strict guidelines for the operators of exhibition facilities and regular inspections during the hours of show operations, exhibit hall operators will not attempt to police their tenants.

All the ingredients for a serious accident or fire are present: a large crowd, excessive quantities of explosive material, and a lack of security. The model fire prevention codes and NFPA 495 adequately address this issue. An adequate program of enforcement by fire officials is needed.

Industrial explosives and explosives actuated tools must be stored and used in accordance with the fire prevention codes. The weight of explosive materials is the determining factor for storage and handling requirements.

Reactive Materials

NOTE
Fertilizer-grade ammonium nitrate is not an explosive material even though it is detonatable under the right conditions.

You will not find provisions for the storage of fertilizer-grade ammonium nitrate in the chapter on explosives in the fire prevention code. Fertilizer-grade ammonium nitrate is not an explosive material even though it is detonatable under the right conditions. Ammonium nitrate and some other substances such as nitromethane and acetaldehyde are classified as *unstable* (*reactive*) substances by the codes and are regulated as hazardous materials. Once ammonium nitrate has been formulated into a blasting agent, the provisions of the explosives chapters apply. Until that time, it is an unstable (reactive) substance and regulated as such. Storage in other than Use Group H buildings is limited to the exempt quantities per building or control area as discussed in Chapter 12.

FIREWORKS

NOTE
An estimated 10,000 persons are injured each year as a result of fireworks injuries, and the fire loss and associated medical costs are in the millions of dollars.

Fireworks is a broad term that can include everything from sparklers to multiple-charge aerial shells fired from mortars. DOT 27 CFR classifies *common* and *special* fireworks and gives examples of each. An estimated 10,000 persons are injured each year as a result of fireworks injuries. Fire loss and associated medical costs are in the millions of dollars.[12]

Why does the American public stand for such a public safety menace? The answer is simple. To the American public, fireworks are every bit as important to the Fourth of July celebration as evergreen trees are to Christmas. There may not be a religious significance attached to the Fourth, but fireworks are a tradition that we hold dear.

Fireworks ordinances generally fall into two categories: those regulating common or consumer fireworks and those regulating aerial and large ground displays. Many jurisdictions have fireworks ordinances separate and distinct from the fire prevention code, which may ban some or all types of common fireworks. Small explosive firecrackers commonly called *lady fingers* are legal common fireworks according to DOT 27 CFR. Many jurisdictions however, ban any firework that explodes, acts as a projectile, or rises vertically in the air.

Fireworks Displays

All model fire prevention codes require permits for public fireworks displays. Depending on the code, application must be made between 10 and 14 days prior to the event. The *UFC* specifically requires the submission of a site plan as part of the permit application. The *BNFPC* and *SFPC* both reference NFPA 1123 *Public Display of Fireworks*, which requires a site plan and additional information.

Permits Permits for public fireworks displays should not be issued without a careful review of the following information:

- Detailed site plan showing all structures, roadways and utilities, spectator and parking areas, and firing and fallout areas
- List and size of all fireworks to be used in the display, manufacturer, supplier, and contact telephone number
- Direction and angle of discharge
- Firing method, mortar size, and construction (see Figure 14-8)
- Name, age, and credentials of pyrotechnician in charge
- Name, age, and credentials of all helpers

Figure 14-8 *Shells must be sorted by size for loading in the appropriate mortar. (Courtesy of Lionel Duckwitz.)*

- Method of crowd and site access control and number of personnel
- Fireworks arrival date, storage, and security arrangements
- Method of two-way communications between operators and security personnel
- Arrangements for misfires and unspent shells
- Arrangements for range cleanup and inspection immediately after the display and at first light
- Bond or insurance certification

■ NOTE
The specified clearance distances should be considered the bare minimum and should never be relaxed.

Minimum clearances dependent on shell diameter are specified in the *UFC* and NFPA 1123. The specified distances should be considered the *bare minimum*, and should never be relaxed. Use of the specified distances will not prevent the fallout of debris and ash outside of the fallout zone. Wind conditions aloft may also be different than conditions at ground level and can carry debris some distance.

Firing the Display Firing the display must be a coordinated effort between the pyrotechnician in charge, his helpers, and site security personnel. The firing area must be secured as shown in Figure 14-9. If at any time lapses in security permit unauthorized persons to enter the fallout area, the fire inspector is obligated to *stop*

Figure 14-9 *Firing areas must be secured to prevent unauthorized persons from entering. (Courtesy of Lionel Duckwitz.)*

■ NOTE
The *UFC* and NFPA 1123 contain basically the same language with the word "shall" indicating that stopping the display is not an option.

the display immediately. The *UFC* and NFPA 1123 contain basically the same language with the word "shall" indicating that stopping the display is not an option.

When shells do not fire, they must be left in the mortar and undisturbed for at least 15 minutes. The mortar should then be flooded with water and left at least 5 more minutes. *After* the display has ended the shell should be placed in a bucket of water and picked up for transport by the supplier. Malfunctioning shells should never be repaired.

■ NOTE
The thorough inspection of the fallout area immediately after the show and then at first light is critical.

The thorough inspection of the fallout area immediately after the show and then at first light is critical. Live, unfired projectiles are commonly found in the fallout area. In the hands of curious children, these can be deadly. They should be treated as misfires and stored immersed in water for pick up by the supplier.

SEIZURE AND DESTRUCTION OF EXPLOSIVE MATERIALS AND FIREWORKS

The model fire prevention codes provide for the seizure and destruction of explosive materials and fireworks under certain conditions. Confiscation generally gets the attention of even the most uncooperative of code violators. Think before you act. Confiscating fireworks can lead to storage and disposal problems (see Figure 14-10). You and the government are obligated to provide the same safe and

Figure 14-10 *Confiscated fireworks.*

code-compliant storage facilities and employ the same safe disposal procedures that the code requires for everyone else. If you seize the dynamite illegally stored in the back of a truck parked in a motel parking lot and then illegally store it in a shed at the fire training center, what have you accomplished? Seizure is not always the preferred method of handling the problem.

Summary

Explosive materials are stored and routinely used in nearly every jurisdiction. Nowhere within the inspection and enforcement process is the inspector's ability to rapidly assess a situation and order appropriate actions more important. Explosive materials are unforgiving and second chances are few.

Inspectors must realize that the difference between an expert in blasting and explosives code enforcement and an expert in blasting and explosives handling is significant. Inspectors should ensure compliance with the fire prevention code and not attempt to serve as operational consultants.

Review Questions

1. Deflagration occurs at ______________ speed.
2. Detonation occurs at ______________ speed.
3. List four explosive materials.
   1. ______________
   2. ______________
   3. ______________
   4. ______________
5. Explosives are stored within structures called ______________.
6. A formula used to determine the maximum charge weight per delay of explosive materials dependent on distance, without damage to surrounding structures is called a ______________.
7. ANFO, or ammonium nitrate and fuel oil mixtures are ______________.
8. The most effective type of stemming material is ______________.
9. At aerial firework displays, fallout areas are established based on shell ______________.
10. When aerial firework shells do not fire, they should be left undisturbed for _____ minutes, then ______________.

Discussion Question

1. You have been summoned to the chief's office to explain the rash of blasting complaints that have inundated the office staff over the past two days. A large vacant tract formerly used by the youth soccer league for a playing field is being developed into low-cost apartments. Neighboring homeowners complain that the blasting operation is shaking their houses and destroying their foundations.
   a. What code provisions are in place to protect the homeowners?
   b. What code provisions protect the interest of the contractor and developer?
   c. What should be your role in the process?

Hazardous Assembly Occupancies

Learning Objectives

Upon completion of this chapter, you should be able to:

- Describe the hazards associated with heliports and helistops on buildings, and how the model codes address them.
- Describe the hazards that make airport terminal buildings require special fire protection features.
- Describe the hazards associated with bowling pin and lane refinishing.
- List five general safety provisions for tents and air-supported structures.

A list of fire incidents in which there was large loss of life includes many that occurred in which large numbers of people were assembled. Assembly occupancies include those buildings where people gather for civic, social, and religious functions, for entertainment, and awaiting transportation. Fires like the Beverly Hills Supper Club in 1977 with 165 fatalities and the Happy Land Social Club in 1990 with 87 fatalities immediately come to mind. However, other assembly occupancies are not immune to the effects of fire. There were 168 fatalities in the Ringling Brothers and Barnum and Baily Circus Tent Fire in 1944. A gas explosion at the Indiana State Fairgrounds Coliseum in 1963 claimed 74 lives and in April of 1996, a fire at the Dusseldorf Airport terminal killed 17 people.

The model fire prevention codes have specific requirements that address special assembly occupancies such as tents, air terminals, and bowling alleys. These requirements extend the level of protection afforded by the model building codes, by regulating the processes and activities within them. Limits on the storage and use of flammable and combustible materials, prohibitions against the use of open flames and heat-producing appliances, and requirements for permits and routine inspections are all included in the model fire prevention codes.

AIR TRANSPORTATION FACILITIES

After the Dusseldorf Airport fire, NFPA's chief fire investigator was quoted as saying "You just don't get structural airport fires like this. You definitely don't get fatalities like this."[1] His comments echoed the feeling of many fire and safety professionals around the world. Millions of people move through airports around the world. Very rarely do fire incidents that do not involve aircraft or aircraft fueling endanger the public.

The model fire prevention codes have requirements for airports, **heliports** and **helistops**, and aircraft hangars. The model building codes have construction requirements that pertain to all three. They also include specific requirements relating to helicopter landing areas located on buildings. Unless specifically designed for the purpose, few buildings will accommodate helicopter landings.

heliport
an area where helicopters take off and land from the ground or water, or from a building; includes areas for storage, maintenance, and refueling

helistop
same as heliport, except without facilities for storage, maintenance, or refueling

NOTE
Unless specifically designed for the purpose, few buildings will accommodate helicopter landings.

Heliports and Helistops

Helicopters take off and land from the ground, from buildings, and from bodies of water at heliports and helistops. Heliports include areas for storage and maintenance and for refueling operations. Helistops are strictly for take off and landing, with the loading and unloading of passengers or freight.

The increased use of medivac helicopters has led to the establishment of heliports and helistops at many medical facilities. Because takeoff and landing operations account for the majority of helicopter accidents,[2] special building code provisions for heliports and helistops located on structures are crucial to the safety of the occupants. In the case of hospital heliports and helistops, the occu-

■ NOTE
Because takeoff and landing operations account for the majority of helicopter accidents, special buildings code provisions for helicopters and helistops located on structures are crucial to the safety of the occupants.

pants of the building may be incapable of evacuation. Building code requirements include structural provisions for the landing pad, special exit provisions, and fire protection requirements.

Fire prevention code requirements address minimum clearances, fire extinguishing equipment, and general fire safety precautions. Permit requirements in the model fire prevention codes are designed to give the fire official notice of the establishment of these facilities and ensure they are inspected with regular frequency.

Airports

Airports are generally large noncombustible structures. Fuel loads within the passenger terminals are limited. Aircraft service and repair areas are a different story. Fire prevention code requirements regarding passenger safety during loading and unloading and refueling operations specifically address control by airline and airport service personnel. Airline and airport personnel must ensure that no smoking is permitted during these times and must ensure that passengers proceed immediately from the terminal to the aircraft. Refueling with passengers on board is only permitted if airline personnel are on board to ensure that passengers evacuate on the other side of the aircraft in the event of an incident.

Aircraft Hangars

Aircraft hangars are classed by the model building codes as moderate hazard storage facilities. The *UBC* classifies hangars where major engine overhauls are conducted as Group H Division 5. Aircraft hangar fires are often high fire losses due to the value of the aircraft. The large quantities of flammable liquids and solvents and the ignition potential from machinery and tools necessitate strict fire prevention practices. No smoking should be strictly enforced. Fire protection equipment and systems should be scrupulously maintained. Housekeeping should be meticulous.

BOWLING ALLEYS

Two hazardous processes in bowling alleys created the need for special chapters or articles in the model fire prevention codes. Bowling pin and floor refinishing using flammable finishes involves two hazardous conditions. Floor sanding, in which power sanding equipment removes the old floor finish and wood surface creates a large volume of combustible dust. Floor and pin refinishing using flammable finishes creates hazardous vapor–air mixtures. In the case of floor finishing, the entire building is often filled with flammable vapors.

Permits are required by the model fire prevention codes for floor or pin refinishing using flammable finishes. The permit requirements allow the fire offi-

■ **NOTE**
In the case of floor finishing, the entire building is often filled with flammable vapors.

cial to inspect the facility during operations to ensure compliance with the code. In addition to general fire safety precautions:

- Floor resurfacing operations are prohibited while the facility is open to the public. This includes both sanding and the application of finishes.
- All heating, cooling, and ventilating systems that recirculate air must be shut down and not operated within 1 hour of application.
- All motors and ignition sources must be shut down and not operated within 1 hour. Smoking is also prohibited.
- Pin refinishing and spraying must be conducted within a rated room as specified by the model building codes.
- Waste material and dust must be stored in approved containers.

The development of high quality latex floor finishes and synthetic bowling pins have greatly reduced the use of flammable finishes in bowling alleys.

TENTS AND AIR-SUPPORTED STRUCTURES

The SFPC defines a tent as *"an enclosure constructed of canvas or other pliable material . . ."* The *IFC* and *UFC* have similar definitions, then adds one for canopy, which is similar, but *"without sidewalls for at least 75% of the perimeter." BNFPC* does not define tent at all, but a canopy in the *BNBC* is *"an architectural projection attached to a building."* Neither the *BNFPC* nor the *SFPC* contain definitions for air-supported structures, but the *UFC* defines temporary membrane structures as an *"air-inflated, air-supported cable or frame covered structure . . . ."* Permit requirements are just as varied, as shown in Table 15-1. Regardless of which code you are using, certain provisions apply to all tents and air-supported structures:

- They must be structurally sound and in good repair. Anchoring and inflation systems must be installed in accordance with the building code.
- Fabric coverings must be of flame-retardant material or be treated in an approved manner. Certification must be provided.
- Seating plans must be submitted for approval with construction documents during permit application.

Table 15-1 *Permit requirements.*

	IFC	BNFPC	SFPC	UFC
Tents	200 ft^2	900 ft^2	120 ft^2	200 ft^2
Canopies	400 ft^2	n/a	n/a	400 ft^2
Air-supported or temporary membrane	200 ft^2	900 ft^2	120 ft^2	200 ft^2

- Open flame devices or unapproved heating devices are prohibited inside or within 20 feet of the tent or air-supported structures.
- Combustible materials such as hay are prohibited except as needed for the feeding of animals. Sawdust or shavings must be kept damp.
- Exit lights and exit illumination are required.
- Heating and cooking equipment must be approved.
- Vegetation and trash or waste material must be kept cleared from around the tent or air-supported structure.
- Approved fire extinguishers must be provided.

Problems regarding the application of the fire prevention code regarding tents can have considerable political impact. Citizens who are unaware of permit requirements or regulations are not normally at their best when a fire inspector arrives during their daughter's wedding reception. The effect can be enhanced when the inspector orders the caterer to stop cooking.

■ NOTE
The political impact of one inspector's actions can affect the whole fire department.

Permit requirements and code provisions should be given to the customer during the permit application process. Tent suppliers within the jurisdiction should be well versed in exactly what the fire prevention bureau requires and expects. The political impact of one inspector's actions can affect the whole fire department during the next budget cycle. I am not advocating lax enforcement. *I am advocating no surprises.*

Summary

Fires in special assembly occupancies combine the hazards of high occupant load with the unique fire potential of the specific occupancy. Airports would be less hazardous than large mercantile occupancies (less combustible contents), if not for the aircraft and fuel that surround them. A wedding reception at the Elks Lodge is not particularly hazardous. The same event in a tent can create a multitude of hazardous conditions. The model fire prevention codes require permits for certain processes to enable the fire official to effectively regulate them. Directives can be given at the time of permit application. Inspections will verify that the process or occupancy is in compliance with the code.

Review Questions

1. List three building code requirements for heliports and helistops located on buildings.
   1. ____________
   2. ____________
   3. ____________
2. Under what circumstances are passengers allowed to board an aircraft that is being refueled? ____________
3. The major hazard associated with aircraft hangars is ____________.
4. What are the two major hazards associated with bowling alleys?
   1. ____________
   2. ____________
5. When resurfacing bowling lanes with flammable floor finish, all motors, fans, and ignition sources must be shut down for at least ____________.
6. Tent fabric must be made of ____________ or be ____________ in an approved manner.
7. Under what circumstances are combustible materials such as hay and sawdust permitted inside of tents? ____________
8. The majority of helicopter accidents occur during ____________.
9. Who is responsible to ensure that no smoking rules are observed during aircraft passenger loading and unloading? ____________
10. Open flame devices or unapproved heating devices are prohibited inside or within ____________ of a tent or air-supported structure.

Discussion Question

1. You have been contacted by the manager of the local bowling alley. She is refinishing the floors with a latex floor finish starting Saturday night at midnight. She will be sanding and finishing half the lanes Saturday night and reopening on Sunday. She plans on

sanding the other half Sunday during the day and will wait to apply the floor finish until Sunday night after closing.

a. Does she need a permit to conduct this operation?

b. Does her plan comply with the provisions of your model fire prevention code?

Other Storage and Processing Occupancies

Learning Objectives

Upon completion of this chapter, you should be able to:

- Discuss the value of corporate safety programs in industrial occupancies.
- List three of the five main causes and contributing factors involving storage facility fires.
- Explain the reasons for the proliferation of waste material and rubbish handling facilities.
- Describe how the provisions of the model fire prevention codes attempt to reduce fire risk at lumberyards.

■ NOTE

During 1995, eighteen fires with losses of 10 million dollars or more were reported, and sixteen of those were in industrial type occupancies.

In NFPA's annual summary on large loss fires during 1995, eighteen fires with losses of 10 million dollars or more were reported. Of those, sixteen were in industrial type occupancies and accounted for more than $1,003,200,000.[1] One fire in a textile finishing mill in Massachusetts resulted in a direct loss of $500 million. The actual loss of such fires is much greater than the loss of the building and contents. The loss of jobs and buying power has a catastrophic effect on the entire community, and the effects can be far reaching. Factories across the country can be idled by the abrupt loss of parts from a supplier. In one way or another, these fires affect all of us, either directly, or through increased costs for consumer goods, insurance, or consumer credit.

Investigation reports of fires in industrial occupancies often identify factors that contributed to the large loss. Inadequate fire protection system design, improper maintenance, and change in hazard classification are commonly noted, as are the storage of incompatible materials, temporary storage within aisles, and increased height of storage.[2]

■ NOTE

An effective inspection program that includes the education of facility managers can effectively reduce the likelihood of a large fire loss.

All of these issues are addressed by the model fire prevention codes. An effective inspection program that includes the education of facility managers can significantly reduce the likelihood of a large fire loss. Similar fires in chicken processing plants in Hamlet, North Carolina, and North Little Rock, Arkansas, clearly underscore the value of fire prevention practices.

The fires occurred in chicken processing plants approximately 3 months apart, in 1991. The Tyson plant in North Little Rock had a corporate fire prevention program in place that included the appointment of a fire safety director, mandated fire drills, established a fire brigade, and restricted combustible storage in hazardous areas. The fire, which occurred when a high pressure hydraulic line failed and atomized hydraulic fluid was ignited, caused 8 million dollars in damage and resulted in the plant being closed 13 weeks. All 115 employees evacuated within 3 minutes, and all were accounted for using the procedures practiced during routine fire drills. Fire brigade personnel with self-contained breathing apparatus (SCBA) led suppression crews to the fire area.[3]

In contrast, the fire at Imperial Foods in Hamlet, North Carolina, resulted in twenty-five fatalities and fifty-four injuries among the approximately ninety persons working at the plant at the time of the fire. As in the North Little Rock fire, a ruptured hydraulic line in the cooking area led to the fire. Locked and blocked exit routes and the lack of an evacuation plan led to widespread confusion. One employee perished when she became wedged between a compactor seal and the building wall. Twelve employees took refuge in a walk-in cooler and succumbed to the fire.

Chief Fuller of the Hamlet Fire Department was quoted as saying that the entire incident centered around one problem—the lack of enforcing existing codes. The Hamlet Fire Department had no code enforcement, and the North Carolina Department of Occupational Safety and Health had never inspected the facility during its 11 years in operation.[4]

STORAGE FACILITIES

The main causes or contributing factors in fires in storage occupancies are bad housekeeping, industrial trucks (forklifts), welding and cutting, smoking, and arson.[5] All are addressed by specific fire prevention code provisions.

■ **NOTE**
The model codes contain significantly different requirements for storage facilities, depending on the hazards posed by the commodities stored.

In Chapter 3 we learned that the model codes contain significantly different requirements for storage facilities, depending on the hazards posed by the commodities stored. Warehouses could be classified as low, moderate, or high hazard, and even within those classifications NFPA 13 and UBC 9-1 have different sprinkler requirements depending on the commodities stored. Inspections of storage facilities should include verification that the facility has not changed to a more hazardous use. A check of the hydraulic nameplate affixed to the sprinkler riser will indicate the original hazard classification based on the design density (see Figure 16-1).

■ **NOTE**
Inspections of storage facilities should include verification that the facility has not changed to a more hazardous use.

Idle wooden or plastic pallets create a significant fire load. A maximum storage height of 6 feet is all that *ordinary hazard* sprinkler protection can effectively protect.[6] Storage heights over 6 feet require an increased density.

Warehouse facilities should have designated smoking areas (even if they are outside they should be designated as such), with suitable ashtrays. Waste containers should be substantial and should be emptied on a regular basis. Spare cylinders for forklifts should be secure, protected against damage, and comply with quantity restrictions of the fire prevention code.

Figure 16-1 *Improper storage practices can severely compromise sprinkler protection. (Courtesy of Lionel Duckwitz.)*

■ NOTE
Idle wooden or plastic pallets create a significant fire load; a maximum storage height of 6 feet is all that ordinary hazard sprinkler protection can effectively protect.

The storage of matches, flammable liquids, aerosols, and other hazardous items within general purpose warehouses entails significant restrictions and safeguards. The code provisions from the applicable chapters or articles of your model fire prevention code and the referenced standards must be enforced.

WASTE MATERIAL HANDLING FACILITIES

Developments since the late 1980s have changed the way we feel about trash. When our landfills began to fill up and the public began to support environmental issues, laws were passed to encourage and in some cases mandate recycling. That trash that fire inspectors have battled for years, relegating it to covered metal containers to be hauled quickly away suddenly became a valuable resource. Waste material handling plants have become big business (see Figure 16-2).

■ NOTE
When governments and major newspapers announced policies of only printing on paper that contained recycled material, trash became valuable.

When governments and major newspapers announced policies of only printing on paper that contained recycled material, trash became valuable. Then in 1994, the United States Supreme Court ruled that private companies must be allowed to compete with local governments in establishing transfer stations. Transfer stations are facilities where small amounts of refuse are dumped, compacted, and shipped to distant landfills in distant areas.

Permits are required by all the model fire prevention codes for the operation of waste material and rubbish handling facilities. This includes recycling facilities where newspapers are separated and baled for shipping.

Figure 16-2 *Waste material handling.*

■ NOTE
The illegal storage of waste tires in buildings poses an extra challenge; the obvious code violations are compounded by concerns for fire fighter safety.

Work closely with your building official to ensure that building use group, sprinkler system design, and building separation issues are addressed. The establishment of these facilities can be particularly political and contentious. Extra attention to the details is well worth the effort.

The illegal storage of waste tires in buildings poses an extra challenge. The obvious code violations are compounded by concerns for firefighter safety in the event of a fire, and in many cases, the criminal nature of the storage. Illegal tire dumping is a felony in some jurisdictions. Close coordination between governmental agencies is necessary. The safety of firefighters and the public must be the primary objective.

LUMBERYARDS AND WOODWORKING PLANTS

■ NOTE
The sheer volume of the exposed surface area of the lumber piles is conducive for radiant heat transfer between piles.

Fires in lumberyards are especially challenging for firefighting forces. Lumber is stacked with spacers (see Figure 16-3) to enable forklift operators to quickly remove a large number of boards at once. These spacers create flue spaces in the event of a fire. The sheer volume of the exposed surface area of the lumber piles is conducive for radiant heat transfer between piles. Emergency apparatus access is critical because large quantities of water are needed and are applied with deck guns and elevated streams.

Figure 16-3 *Lumber is stacked to facilitate moving with forklifts. This creates horizontal flue spaces.*

The model fire prevention codes attempt to minimize the hazards associated with lumber storage by:

- Limiting piles to 150,000 cubic feet.
- Limiting lumber piles to 20 feet in height.
- Requiring driveways on 50 foot by 150 foot grids.
- Requiring a 6-foot security fence around the storage.
- Requiring an adequate water supply and fire protection equipment.
- Prohibiting smoking.
- Requiring the removal of combustible waste and vegetation.

board foot
a measurement used for lumber equalling 144 cubic inches (12 inches by 12 inches by 1 inch)

Permit requirements for lumberyards are based on the number of **board feet** (fbm) stored at the site, not cubic feet. To determine whether a facility needs a permit while you are in the field, use the following rule of thumb. Since:

A board foot is 1 foot by 1 foot by 1 inch thick, and

A cubic foot of wood contains 12 board feet (1 foot by 1 foot by 12 inches), then

A lumber pile 20 feet by 20 feet by 20 feet in height = $20 \times 20 \times 240$ (from 20 feet $\times$ 12 inches) = 96,000 fbm.

Add the piles together. Anything bigger than 20 feet by 20 feet by 20 feet requires a permit. Very few facilities will be within a questionable range. If they are in the lumber business, they will have well over 100,000 board feet; if not, they will generally have considerably less.

■ NOTE
Woodworking plants are potentially hazardous due to the quantity of combustible wastes that are generated.

Woodworking plants are potentially hazardous occupancies due to the quantity of combustible wastes that are generated. Large volumes of sawdust and shavings are created by woodworking equipment. Automatic waste removal systems must comply with the requirements of the mechanical code and standards referenced by your model fire prevention code. At a minimum, floors must be swept at the end of each shift and floor sweepings must be removed from the building or stored in closed metal bins.

Inspectors should examine machinery and interior surfaces within the facility for evidence of sawdust accumulation. Sawdust must be removed from interior surfaces including bar joists and rafters. Accumulations may indicate that the waste removal system is not functioning properly. Properly sized and maintained fire extinguishers are especially important in these facilities.

Summary

Large loss fires in industrial and storage facilities have a significant economic impact on society at large. We all pay through increased costs for consumer goods and services, insurance, and consumer credit. Effective fire prevention programs implemented by state and local governments, the insurance industry, and the business community can reduce the risk of fire.

General precautions as well as specific requirements from the fire prevention code must be enforced. Inspectors should verify that a facility has not changed to a more hazardous use and that fixed fire suppression systems are still adequate.

Review Questions

1. List six factors that have been identified as contributing to large losses in industrial occupancies.
   1. ______
   2. ______
   3. ______
   4. ______
   5. ______
   6. ______
2. List five causes or contributing factors to fires in storage occupancies.
   1. ______
   2. ______
   3. ______
   4. ______
   5. ______
3. The original hazard classification of a sprinklered warehouse can be determined by checking the ______, listed on the hydraulic nameplate, located on the sprinkler riser.
4. The maximum height of idle pallet storage that can be protected by an NFPA 13 *ordinary hazard* sprinkler system is ______.
5. In industrial occupancies, smoking should only be permitted ______.
6. List four ways in which the model fire prevention codes address the hazards associated with lumber storage.
   1. ______
   2. ______
   3. ______
   4. ______
7. 1 board foot measures ______ by ______ by ______ thick.
8. Woodworking plants are potentially hazardous occupancies due to ______.
9. At a minimum, floors in woodworking plants should be swept ______.
10. Accumulations of sawdust on bar joists and rafters may indicate that ______.

Discussion Question

1. During an inspection of a fully sprinklered cabinet-making shop of noncombustible/unprotected construction, you find that a small travel trailer has been moved in and is being used as an office. The wheels have been removed, and it has been connected to the building electrical service. The owner sheepishly informs you that he knew you would have a problem with the lack of sprinkler protection in his new office and offers to immediately have it installed.

   a. What code violations exist regarding the trailer?

   b. What would be your actions?

Compressed Gases and Cryogenic Liquids

Learning Objectives

Upon completion of this chapter, you should be able to:

- Describe the four physical states in which gases are commonly stored within containers.
- Define the term *gas*.
- Describe the *Combined Gas Law* and its significance.
- List two fire safety strategies for gas storage as they relate to the Combined Gas Law.
- Explain why LPG container storage requirements are expressed in water gallon capacity.

PHYSICAL PROPERTIES OF GASES

compressed gas
a gas or mixture of gases having an absolute pressure exceeding 40 psi at 70°F, or an absolute pressure exceeding 140 psi at 130°F, or any liquid with a vapor pressure that exceeds 40 psi at 100°F

cryogenic liquid (or fluid)
a refrigerated liquid gas with a boiling point below −130°F (SFPC 94) or below −150°F (UFC 94) or below −200°F (BNFPC 96)

> Solids retain their size and shape. Liquids have a definite size and at least one shape, a flat surface. But gases have neither. No size, no shape and, seemingly, no rhyme or reason.[1]
>
> (James H. Meidl)

It would be hard to imagine our lives without **compressed gases** and **cryogenic liquids**. Compressed gas refrigeration systems cool our buildings and preserve our foods. Anhydrous ammonia is used to fertilize our croplands. Gases are used as anesthetics during surgical procedures, while oxygen supports our breathing during surgery. Steel fabrication, chemical processing, and even fixed fire protection systems depend on gases. Your soft drink at lunch would have been pretty awful without the bubbles provided by carbon dioxide gas.

Like flammable and combustible liquids, the same characteristics that make compressed gases and cryogenic liquids useful can become deadly attributes with misuse or mishandling. High heat output in combustion, high reactivity in chemical processing, extremely low temperatures, and the economy of handling large volumes easily in compact form are attributes that make compressed gases invaluable in our every day lives.

■ NOTE
A gas is any substance that boils at atmospheric pressure and at any temperature between the absolute zero of outer space and approximately 80°F.

The term *gas* refers to a state of matter. For our purposes, a gas is any substance that boils at atmospheric pressure and at any temperature between the absolute zero of outer space and approximately 80°F.[2] Of the ninety-two natural elements, eleven are gases under standard conditions[3]: hydrogen, nitrogen, oxygen, fluorine, chlorine and the noble gases—helium, neon, argon, krypton, xenon, and radon, which are inert. An apparently unlimited number of compounds, such as air, also have similar boiling points and thus exist as gases.

■ NOTE
Good practice and compliance with the code provisions for a nonflammable compressed gas may be totally inadequate for a toxic gas or a gas that is an oxidizer.

Some gases and gas mixtures are flammable, others are toxic; some are reactive, some are inert. Some pose multiple hazards. Some are colorless and odorless, some are heavier than air and will settle and form pockets, waiting for an ignition source. Inspectors must consider the relative hazards based on the physical properties of the gas or gas mixture. Good practice and compliance with the code provisions for a nonflammable compressed gas may be totally inadequate for a toxic gas or a gas that is an oxidizer.

The Physical Laws of Gases

Somewhere in your educational past, an instructor probably explained a series of scientific laws or facts about gases. The physical properties of gases are the basis for the requirements of the model fire prevention codes, so let us review a moment.

> *Boyle's law*[4] states that the volume occupied by a given mass of gas varies inversely with the absolute pressure, if the temperature remains constant.

■ NOTE
The pressure of a gas depends on the volume of the container it is in and the temperature it is exposed to.

Charles's law[5] states that the volume of a given mass of gas is directly proportional to the absolute temperature, if the pressure remains constant.

The *Combined Gas Law*,[6] which combines Boyle's and Charles's laws, states that the pressure of a gas is dependent upon the volume of the container it is in and the temperature it is exposed to.

Gaseous States in Storage

Gases in storage containers exist in one of four states:

- Nonliquefied gases, which are stored under their charged pressure and are entirely gaseous at 68°F.
- Liquefied gases, which are stored under their charged pressure and are partially liquid at 68°F.
- Compressed gases in solution, which are nonliquefied gases dissolved in a solvent.
- Cryogenic liquids or fluids, which are liquefied refrigerated gases with boiling points ranging between –135°F and absolute zero (–460°).

Strategies for Safety

With the physical properties of gases in mind, consider the basic strategies that should be used to ensure the safe storage, use, and handling of compressed and liquified gases:

1. Keep the temperature constant so the pressure does not increase.

   Control combustible construction, combustible storage, and vegetation to limit fire exposure.

   Provide a fixed fire protection system to cool containers.

   Limit quantities in storage and use to limit the exposure potential.
2. Increase the volume of the container if the temperature increases.

   Provide overpressure protection such as pressure relief valves and fusible plugs.
3. Protect against container failure at normal temperatures.

   Require all containers, valves, and fittings to conform to a standard.

   Require periodic hydrostatic testing of containers.

   Require that valves and containers be protected against falling and physical damage (see Figure 17-1).
4. Protect against the specific physical hazards of the gas.

   Provide ventilation, fixed fire protection, and emergency containment.

Figure 17-1 *This CNG fueling facility has vehicular protection and no combustible material in proximity.*

Develop specific code requirements that address flammability, reactivity, toxicity, and special hazards.

■ **NOTE**
All containers must be identified and marked in accordance with Department of Transportation regulations.

CODE REQUIREMENTS FOR GASES

Code requirements applying to gases are found within the model fire prevention codes in various articles and chapters, including Compressed Gases, Cryogenic Liquids or Fluids, Welding and Cutting, Fruit Ripening Processes, Refrigeration, Hazardous Materials, Liquified Petroleum Gas, Natural Gas Systems, and Fumigation. These articles and chapters have referenced standards that range in scope from bulk oxygen systems at health-care facilities to compressed natural gas as a vehicle fuel.

nesting
method of securing compressed gas cylinders in groups, in which each cylinder has a minimum of three points of contact with other cylinders, walls, or bracing

General requirements for all gases include:

- All containers, cylinders, pressure vessels, valves, and piping must be designed, constructed, and maintained in accordance with national standards including ANSI, DOT, and NFPA.
- All containers must be identified and marked in accordance with DOT regulations.
- Containers must be protected against physical damage and falling with adequate barrier protection or bracing, or by **nesting**.

■ NOTE
Containers must be protected against physical damage and falling with adequate barrier protection or bracing, or by nesting.

- Liquefied compressed gas cylinders and containers and cylinders of gas dissolved in solvent must be stored in an upright position.
- Containers must be protected against exposure from fire by maintaining clearance from combustibles and clearing vegetation.
- Adequate separation must be maintained between gas storage and other hazards, including other gases.
- Limits on quantities by building use group and as specified in the fire prevention codes must be maintained.

CONSIDERATIONS FOR SPECIFIC GASES

Liquified Petroleum Gas

■ NOTE
Liquefied petroleum gas (LPG) is not a specific substance, rather, it is a mixture of gases.

Liquified petroleum gas (LPG) is not a specific substance, rather, it is a mixture of gases. LPG is composed primarily of propane and butane with small amounts of propylene, ethane, ethylene, and other hydrocarbons[7] and is never quite the same mixture.

■ NOTE
LPG quantity limits for tanks and containers are specified by the capacity of the vessel in water gallons.

Throughout the codes, quantity limitations of gases are specified in cubic feet of gas. Because the physical properties of the constituent gases that make up LPG are different, as shown in Table 17-1, the standard system is impossible. One hundred cubic feet of LPG today and 100 cubic feet of LPG tomorrow will not be the same amount of gas, because you never have the exact same mixture twice. This problem is remedied by using a constant that never changes. LPG quantity limits for tanks and containers are specified by the capacity of the vessel in water gallons.

vapor density
the ratio of the weight of a given volume of a gas to that of air

Because the **vapor densities** of butane and propane are 2.0 and 1.6, respectively, LPG is heavier than air and any leak normally finds its way to the lowest area or part of the building. Prohibitions within the model fire prevention codes and referenced standards against the use of containers and first stage regulating equipment within buildings or on the roofs or balconies of buildings are based in part on the vapor density of the gas mixture.

■ NOTE
LPG is heavier than air and any leak normally finds its way to the lowest area or part of the building.

Table 17-1 *Properties of LPG constituent gases.*

Gas	Butane	Propane	Ethylene	Propylene
Density at ambient temperature (lb/ft³)	.15537	.11599	.07870	.11044
Weight/gallon	4.865	4.223	4.735	4.343
Specific volume (ft³/lb)	6.3356	8.4515	12.7	8.875

Source: Compressed Gas Association, *Handbook of Compressed Gases* (New York: Van Nostrand Reinhold, 1981), p. 243.

The proliferation of LPG barbecue grills in residential occupancies, especially multifamily occupancies, poses a formidable challenge to fire suppression forces as well as fire prevention bureaus. Cylinder exchange programs, in which consumers return empty containers for full containers have created the hazardous situation of consumers carrying partially filled containers into busy mercantile occupancies seeking information on exchanging them (see Figure 17-2). Fire inspectors should require facilities that offer this service to prominently post warnings on the front entrance that LPG cylinders cannot be brought into the building.

Medical Gases

Medical gases including flammable anesthetics and oxygen must be stored in accordance with the general requirements for all gases, as well as the requirements of NFPA 50, NFPA 99, UFC 74-1, and 74-2. Fixed piping systems cannot be used to distribute flammable medical gases such as anesthetics. Only nonflammable gases may be distributed through fixed piping systems.

Toxic Gases

Toxic gases are severely limited in use and storage. Exempt quantities are such that few facilities other than Use Group H structures will have much around. Re-

Figure 17-2
Consumers bringing LPG cylinders into mercantile occupancies pose a severe hazard. (Courtesy of Duane Perry.)

quirements for gas cabinets, exhausted enclosures, treatment systems, and other safety requirements are extensive.

As we discussed in Chapter 12, employees must be trained in the relative hazards, receive training in handling emergencies, and be fully familiar with breathing apparatus and emergency equipment provided. Every inspection should include a review of training records and inspection of the readiness of breathing apparatus and emergency equipment. As you tour the facility, ask an employee when was the last time he or she had the breathing apparatus on.

NOTE

As you tour a facility, ask an employee when was the last time he or she had a breathing apparatus on.

Fumigation operations in which toxic gases are used to eliminate insect infestations have specific code requirements that include the posting of the property and the stationing of sufficient, capable alert watchmen, with breathing apparatus, to guard against the entry of any person. Inspectors should check the condition of breathing apparatus and verify that the watchmen have communications capability such as a cell phone, or have access to a telephone. They must understand that they are to remain *awake* and *alert*.

Anhydrous Ammonia Anhydrous ammonia is used in agriculture as a fertilizer. It is a key element in explosives manufacturing. It is also used in building refrigeration systems and in blueprinting. It is flammable, but its flammable range is extremely high: a 16 percent concentration with air is needed to reach the lower flammable limit.[8] It is a corrosive gas, so quantities are limited to the exempt amounts in all use groups, except the appropriate high hazard prescribed by your model building code. Leaks from blueprinting equipment illegally installed within business use buildings pose a significant risk to the occupants. Anhydrous ammonia is particularly hazardous to those with respiratory problems.

Ethylene Gas Fruit and crop ripening with ethylene gas poses two hazards. Ethylene is flammable and has an anesthetic effect. The heating of buildings used for fruit and crop ripening must be through indirect methods such as steam or hot water, or with sealed listed heating equipment.

Acetylene Welding and cutting using acetylene is only permitted in areas approved for the purpose, such as factory/industrial occupancies or areas that have been made safe for such operations. Permits required by the model fire prevention codes are to cut and weld in areas other than factory/industrial type areas. The permit requirement enables the fire official to make the provisions of the code clear to the contractor. It also enables the fire official to inform the contractor that if he does not follow the rules, his permit will be revoked and he is out of business in the jurisdiction. Welding and cutting within occupied buildings is dangerous business. The model codes recognize the relative hazards. The SFPC requires the area to be free of combustibles and the floor swept clean for a radius of 35 feet. A capable fire watcher with fire extinguishing capability must be posted and remain

■ NOTE **Acetylene cylinders must be maintained in the upright position, which is considered to be within 45 degrees of vertical.**

for at least 30 minutes after operations are complete. Acetylene cylinders should be stored away from oxygen cylinders, and clearances required by the codes should be maintained. Acetylene cylinders must be maintained in the upright position. An upright position is considered to be within 45 degrees of vertical.

Cryogenic Liquids Cryogenic liquids or fluids are gases that are liquefied by cooling them below their boiling points. To put the process into perspective consider this: Gasoline has a boiling point that is somewhere between 100° and 400°F. Its flash point is considerably lower, about –45°F. Methane, the principal component of natural gas (NG) has a boiling point of –289°F. With a boiling point that low, its flash point really is not important. Cool methane below –289° and it becomes a liquid and can be stored and transported easily in small containers. Cryogenic liquids must be refrigerated to maintain them in the liquid state and prevent pressure buildups.

Cryogenic liquids present hazards in three basic ways. First is the hazard posed by the substance itself. Flammability, toxicity, reactivity, or action as an oxidizer are all possibilities. Second, cryogenic liquids have incredible liquid-to-vapor expansion ratios. The liquid-to-gas ratio of cryogenic nitrogen is 697 to 1![9] Gases that are not toxic can quickly kill by asphyxiation. Third is the hazard posed by the extreme cold of the liquid. Injuries to the human tissues from the extreme cold resemble burn injuries. The tissue is destroyed.

Summary

Matter exists in three states—solid, liquid, and gaseous. Gases assume the shape of whatever container or space they are introduced into. The Combined Gas Law, based on Boyle's and Charles's laws, states that the pressure a gas exerts upon its container is dependent upon the volume of the container and the heat that the gas is exposed to. The model fire prevention codes use storage limitations, protection from exposures, and pressure relief to address the hazards posed by gases described by these laws of physics.

Only gases are totally capable and ready fuels (those that are flammable). They require no preheating. The only restrictions are the amount of oxidizer available and the heat required for ignition.

Gases are stored in four basic physical states: nonliquefied, liquefied, compressed gases in solution, and cryogenic liquids or fluids.

Review Questions

1. List the four states in which gases are stored in containers.
   1. ______________________
   2. ______________________
   3. ______________________
   4. ______________________
2. The Combined Gas Law states:

   ____
3. List four safety strategies used in dealing with compressed and liquefied gases.
   1. ______________________
   2. ______________________
   3. ______________________
   4. ______________________
4. A method of securing compressed gas cylinders in groups in which each cylinder has a minimum of three point of contact with other cylinders, walls, or bracing is called

   ______________.
5. Liquefied compressed gas cylinders and cylinders of gas dissolved in solvent must be stored ______________.
6. Because LPG is a mixture of gases with different physical properties, tank and container size is expressed in terms of

   ______________.
7. Gases with vapor densities greater than 1 are ______________ than air.
8. Fire inspectors should require facilities that offer LPG cylinder exchange service to ______________.
9. Watchmen at the scene of fumigation operations must have ______________ and ______________.
10. ______________ are gases that are liquefied by cooling them below their boiling points.

Discussion Questions

1. Why are the threads on flammable compressed gas cylinders designed to require attachment using a counterclockwise motion?
2. You have arrived on the site of a house to be fumigated with a toxic gas to eliminate an insect infestation. The contractor has applied for a permit as required by the fire prevention code. He advises that his employee who will act as the watchman have access to a pay phone on the next block and a self-contained breathing apparatus (SCBA). Upon questioning the job foreman, you discover that the watchman plans to spend the night in a small tent in the front yard of the residence. When you ask the watchman to allow you to examine the SCBA you realize that he does not speak English.
   a. Does this operation meet the requirements of your fire prevention code?
   b. Should the permit be issued?

Pesticides and Other Health Hazards

Learning Objectives

Upon completion of this chapter, you should be able to:

- Identify the signal words required for all Class I and Class II pesticides.
- List the three categories of pesticides.
- Describe the multiple hazards posed by many pesticides.
- Describe the code provisions that ensure employee and firefighter awareness of the presence of toxic materials.
- Describe the difference between a *toxic material* and a *health hazard material*.

PESTICIDES

■ **NOTE**
Class I and II pesticides are identified by the signal words "POISON," "DANGER–POISON," or "DANGER."

health hazard material
a material that affects the target organs of the body including the liver, kidneys, central nervous system, reproductive system, and circulatory system

The insect and rodent damage to stored grain and foods each year exceeds $500 million in the United States alone.[1] Pesticides are a part of our everyday lives. We all experience low-dose exposures to pesticides on an almost daily basis, but most *reported* injuries are to farmers, loggers, and migrant farm workers.[2] Pesticides are in almost every household and are sold in nearly every grocery, drug store, and hardware store. Class I and Class II pesticides are covered by specific *pesticide storage and display* provisions in the model fire prevention codes. Class I and II pesticides are identified by the signal words "POISON," "DANGER—POISON," or "DANGER" printed on the containers in accordance with the Federal Insecticide, Fungicide and Rodenticide Act (FIFRA). If the signal words are not on the container, the material is not covered by the pesticide provisions. It is regulated as a **health hazard material**.

■ **NOTE**
Accidental poisoning can occur by absorption through the skin and tissues, ingestion, or inhalation.

Pesticides are divided into three groups: rodenticides, herbicides, and insecticides. All are designed to kill and can kill or injure people, pets, and wildlife as effectively as they kill harmful insects, rodents, and plant life. Accidental poisoning can occur by absorption through the skin and tissues, ingestion, or inhalation. When exposed to fire, pesticides can be carried by runoff from hose lines and sprinkler systems and carried with smoke and fire gases, exposing persons both inside and outside the structure.

■ **NOTE**
Many pesticides pose multiple health and physical hazards, and are covered by multiple chapters or articles of the fire prevention codes.

Many pesticides pose multiple health and physical hazards, and are covered by multiple chapters or articles of the fire prevention codes. Pesticides are often also classified as *flammable* or *combustible liquids*, *oxidizers* and *toxic* and *highly toxic liquids*, *solids*, and *compressed gases*. In addition to the fire prevention code requirements that pertain to all flammable and combustible liquids, oxidizers and all toxic and highly toxic solids, liquids, and compressed gases, pesticide storage and handling must comply with the following:

- All containers and packages must be marked.
- Pesticide storage is limited to the first floor, with direct access to the outside.
- Storage rooms must be marked in accordance with the NFPA 704 system or in accordance with UFC Standard 79-3.
- Incompatible materials must be effectively segregated.
- Pesticides cannot be stored in the same area with ammonium nitrate fertilizer.
- Storage areas must be designed to contain hose stream runoff.
- Storage and display shelving must be stable and, where practical, secured to the wall.
- Retail displays must conform to maximum square foot densities, maximum container sizes, and maximum display heights.

When calculating quantities to determine conformance with exempt quantity requirements for use groups or the need for permits, remember that these materials may pose multiple hazards. A quantity of a pesticide that is a *highly toxic liquid* and a *combustible liquid* will be counted twice and added to the inventory under *each* category.

TOXIC MATERIALS, IRRITANTS, SENSITIZERS, AND HEALTH HAZARD MATERIALS

irritant
a noncorrosive chemical that has a reversible inflammatory effect on living tissue at the point of skin contact

sensitizer
a chemical that causes many humans or animals to develop allergic reactions after repeated contact

Irritants, **sensitizers**, and health hazard materials are identified in the model fire prevention codes and are *those materials that do not meet the definition of toxic or highly toxic materials yet pose a danger to humans and animals.* The code definitions for toxic and highly toxic solids, liquids, and gases are derived from federal hazardous materials regulations promulgated by the Department of Labor and are expressed in terms of LD50 and LC50. These refer to *doses* (LD50) by skin absorption or oral ingestion that proved fatal within 1 hour to 50 percent of the laboratory animals, or *concentrations* (LC50) in air that had the same results.

NOTIFICATION, WARNING, AND EMPLOYEE TRAINING

■ NOTE

MSDS sheets for all hazardous materials must be on site regardless of quantity.

All areas where hazardous materials are stored, handled, or used have specific requirements for identification with NFPA 704 or UFC 79-3. MSDS sheets for all hazardous materials must be on site regardless of quantity. Repository containers for fire department use are required over certain thresholds. The requirement for MSDS sheets at buildings below these thresholds is for employee and public safety in the event of a spill or accidental poisoning.

Employees who work with hazardous materials are required to be aware of the potential dangers and be trained in appropriate response to accidents or emergencies. Every inspection should include verification that employee training programs are being conducted as required. Like all records required by the model fire prevention codes, if you wait for the accident or fire to ask for them, they will not be there!

Summary

Toxic and highly toxic solids, liquids, and gases, irritants, sensitizers, and health hazard materials—all pesticides fall into one of these categories. Most pesticides are also a flammable or combustible liquid, compressed gas, or an oxidizer. Pesticides pose multiple hazards. Each hazard must be addressed through the applicable provisions of the code.

Review Questions

1. Class I and II pesticide can be identified by the signal words: ________, ________, or ________.
2. If the signal words are not on the label of a pesticide, it is regulated by the model fire prevention code as a ________.
3. List the three types of pesticides.
   1. ________
   2. ________
   3. ________
4. Accidental poisoning can occur through ________, ________, or ________.
5. List four storage and handling requirements for Class I and Class II pesticides.
   1. ________
   2. ________
   3. ________
   4. ________
6. List three of the multiple hazards that pesticides may pose.
   1. ________
   2. ________
   3. ________
7. Employees who work with hazardous materials are required to be trained ________.
8. A chemical that is not corrosive and has a reversible inflammatory effect on living tissue is called a(n) ________.
9. LD50 refers to lethal doses by ________.
10. LC50 refers to lethal doses by ________.

Discussion Question

1. During an inspection at a large garden supply center, the manager questions the need to provide training for his employees in the appropriate response to spills, releases, or accidents.
   a. How can you verify that a training program is in place?
   b. Are all employees required to be trained?
   c. Why are some pesticides classified as toxic while others are classified as health hazard materials?

Using Referenced Standards

Learning Objectives

Upon completion of this chapter, you should be able to:

- Describe what is meant by *permissive* code language.
- List three features of the standards referenced by the model fire prevention codes.
- List five organizations that develop standards referenced or recognized by the model fire prevention codes.

REFERENCED STANDARDS

■ NOTE
Within the scope of the cite, a referenced standard is considered part of the code.

In Chapter 1 we discussed code development and adoption. Throughout this text, we have referred to various standards referenced by the model codes. These standards regulate construction, material characteristics and performance, and system installation and performance. Without them, effective enforcement of the model codes is impossible.

One hundred seventy-seven different standards are referenced within the 2000 IFC. They range from standards that regulate the maintenance of water-based extinguishing systems (NFPA 25), to dual-walled underground storage tanks (STI P3). Each of the model codes contains references to specific standards *within the code text.* These references cite specific applications for the standards. Within the scope of the cite, a referenced standard is considered part of the code. Advisory provisions and those considered **permissive**, or discretionary for the code official, are not considered **mandatory**.

permissive
code language that is discretionary and leaves application to the judgment of the enforcing authority

mandatory
code provisions that must be enforced and complied with

Not all guidelines for construction, materials, and equipment can be adopted by the model codes as reference standards. Only those developed through nationally recognized consensus processes are adopted by the model codes. Standards must be performance based and not specify particular products or materials. Provisions must be mandatory, not permissive, and the standard must be readily available.

■ NOTE
Standards must be performance based and not specify particular products or materials.

Specific editions are referenced by the model codes and listed by year. Code officials may only enforce the provisions of the edition referenced by the model code, or as amended by the adopting ordinance.

PURCHASING REFERENCED STANDARDS

Code books and referenced standards are expensive. Consider them the tools of the trade. Imagine a surveyor without a transit! Without access to the complete set of the standards referenced by your model fire prevention code you do not have the tools to do the job. Every inspector does not need to carry all one hundred or so documents in the back seat of his car. They should, however, carry a fire and building code in the field. Depending on the jurisdiction, copies of frequently used standards should also be carried in the field. Standards for sprinkler system installation, water-based extinguishing system maintenance, and flammable and combustible liquid storage and handling are a few that are frequently needed.

A complete set of all of the referenced documents should be available at the office, for the use of code enforcement personnel as well as citizens who may have questions. The public has a right to view the code documents. They may not understand them, but they have a right to read the specific provisions that apply to their situation.

Code books and referenced standards should never be thrown away. Previous editions should be maintained in order to determine continued compliance with previously approved conditions. Subscription services, which provide continual updates to referenced standards and recommended practices, are a convenient method of purchasing these documents. When updates and replacement pages arrive, *do not throw away the old pages.* You will end up with a newer edition than is referenced in your model code. Although the newest edition can be used for guidance, *you cannot enforce it.* The fire official may accept compliance with the newer standard as meeting the intent and spirit of the code, if a building owner offers or requests to use the standard. Only the provisions of the referenced edition are mandatory.

Copies of the referenced standards are available from the model code organizations and from the standards organizations themselves. Addresses for the organizations are listed within the model codes.

Summary

Like the model building and fire prevention codes that reference them, national standards are developed through a formal consensus process. The standards are referenced within specific code sections and are listed together within a separate chapter or article. Specific editions of the standards are referenced and thus become an enforceable part of the code. Like nationally developed recommended practices, new editions of national standards can be used for guidance but cannot be enforced.

Review Questions

1. Within the scope of the cite, a referenced standard is considered to be ____________________.
2. Advisory provisions of the standards are considered to be ____________________.
3. Provisions that are not discretionary are considered to be ____________________.
4. Only standards developed through _________ are referenced by the model codes.
5. Specific edition of standards are adopted by reference and listed by ________________.
6. Previous editions of codes and standards should be ____________________.
7. Newer editions of standards than those that are referenced by the model codes may be used by fire officials for __________________.
8. May referenced standards specify particular products or materials? ____
9. Using your model building or fire prevention code, list five organizations that develop standards referenced by the codes.
   1. ______________________________
   2. ______________________________
   3. ______________________________
   4. ______________________________
   5. ______________________________
10. Must standards be readily available in order to be adopted by the model codes? _____

Discussion Question

1. NFPA 58, *Standard for the Storage and Handling of Liquefied Petroleum Gases*, is adopted by reference by the *BNFPC* and *SFPC*, and by transcription as *UFC* Standard 82-1 by the *UFC*.
   a. If there is a conflict between the model code and referenced standard, which provision should be applied?
   b. If a condition is not addressed within the model code or within the referenced standard, what are the fire official's sources for guidance?

Equivalents, Conversion Charts, and Related Mathematics

Volume and Capacity Equivalents

U.S. Gallons	Cubic Meters	Liters	Cubic Inches
1	0.003785	3.785	231.0
1.201	0.004545	4.5	277.41
0.004329	0.00001639	0.01639	1.0
7.481	0.02832	28.32	17.28
264.2	1.0	1000.0	61023.0
0.2642	0.001000	1.0	61.023

Cubic Measure

1728 cubic inches	1 cubic foot
27 cubic feet	1 cubic yard
128 cubic feet	1 cord (wood)
40 cubic feet	1 ton (shipping)
2,150.42 cubic inches	1 standard bushel
231 cubic inches	1 gallon

U.S. Standard Gauge

Gauge	Thickness	Decimal
0	5/16	.3125
3	1/4	.2500
7	3/16	.1875
8	11/64	.1718
10	9/64	.1406
12	7/64	.1093
14	5/64	.078

Conversion Tables

Multiply	By	To Obtain
Diam. circle	3.1416	Circumference
Diam. sphere cubed	0.5236	Volume of sphere
Gallons	0.1337	Cubic feet
Gallons	8.33	Pounds of water
Cubic feet of water	62.427	Pounds of water
Water column feet	0.4336	PSI at base
Inches of mercury	0.4912	PSI

Metric Equivalents

1 centimeter	0.3937 inches
1 inch	2.54 centimeters
1 decimeter	3.937 inches
1 decimeter	0.328 feet
1 foot	3.048 decimeters
1 meter	39.37 inches
1 meter	1.0936 yards
1 yard	0.9144 meters
1 kilometer	0.62137 miles
1 mile	1.6093 kilometers

Measure of Volume

1 cubic centimeter	0.061 cubic inches
1 cubic inch	16.39 cubic centimeters
1 cubic decimeter	0.0353 cubic feet
1 cubic foot	28.317 cubic decimeters
1 cubic meter	1.308 cubic yards
1 cubic yard	0.7646 cubic meters
1 liquid ounce	0.029 liters
1 liter	33.814 liquid ounces
1 liter	1.0567 liquid quarts
1 quart liquid	0.9463 liters
1 gallon	3.785 liters

Miscellaneous Information

To find the diameter of a circle, multiply the circumference by .31831.

To find the circumference of a circle, multiply the diameter by 3.1416.

To find the area of a circle, multiply the square of the diameter by .7854.

To find the surface area of a sphere, multiply the square of the diameter by 3.1416.

To find the number of cubic inches in a sphere, multiply the cube of the diameter by .5236.

Doubling the diameter of a pipe or hose increases its capacity four times.

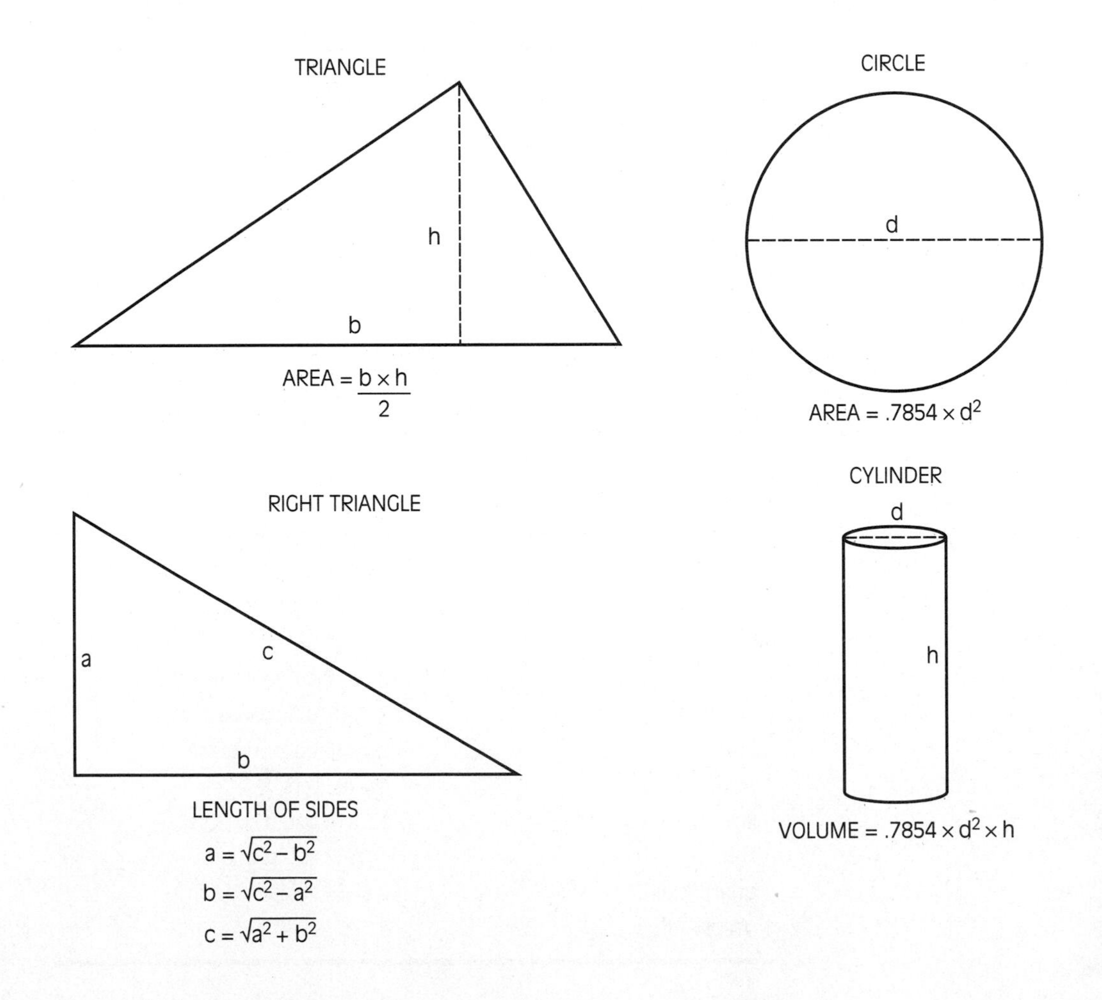

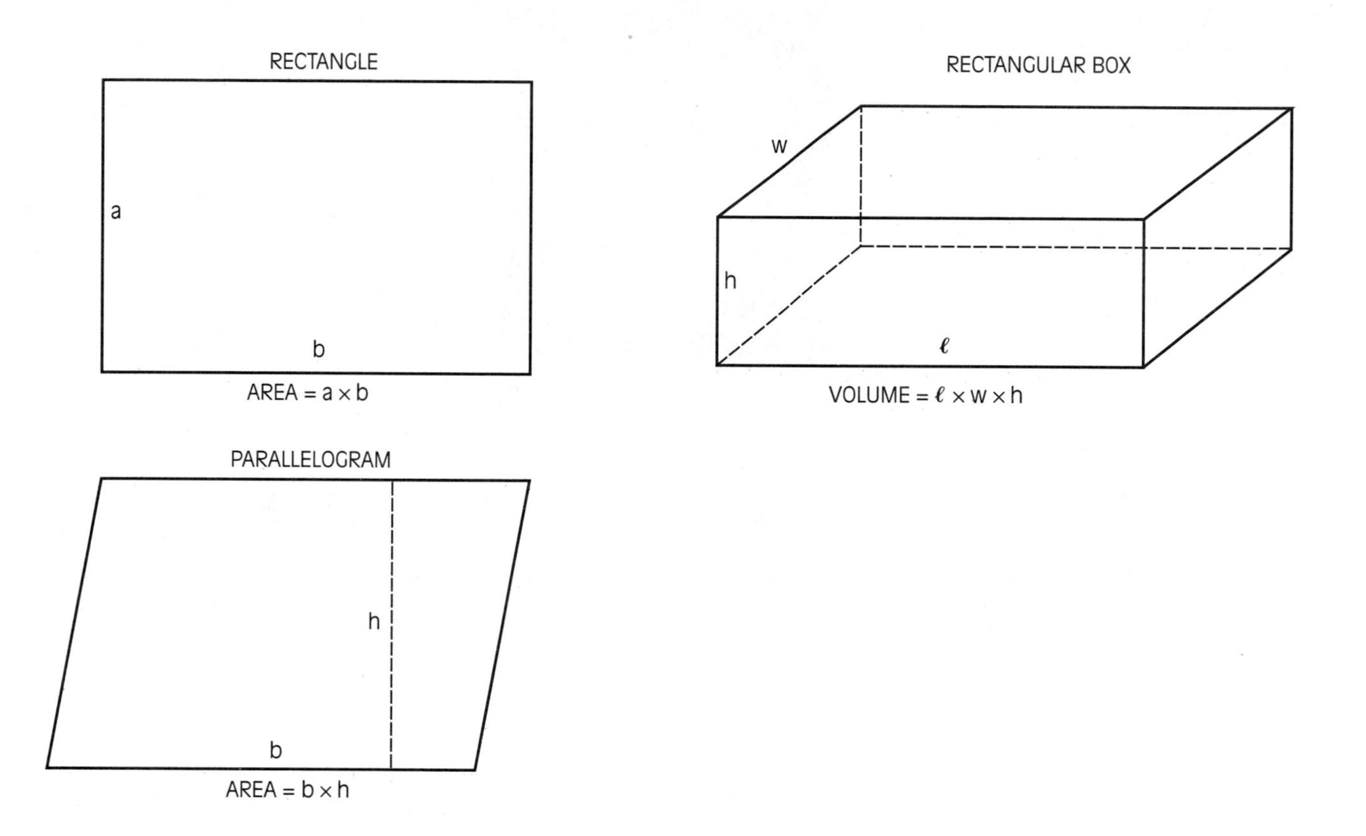
RECTANGLE
a
b
AREA = a × b
RECTANGULAR BOX
w
h
ℓ
VOLUME = ℓ × w × h
PARALLELOGRAM
h
b
AREA = b × h

Numerical Index—2000 National Fire Codes Contents

33	Spray Application Using Flammable or Combustible Materials—1995	Vol 2
34	Dipping and Coating Processes Using Flammable or Combustible Liquids—1995	Vol 2
35	Manufacture of Organic Coatings—1995	Vol 2
36	Solvent Extraction Plants—1993	Vol 2
37	Stationary Combustion Engines and Gas Turbines—1994	Vol 2
40	Storage and Handling of Cellulose Nitrate Motion Picture Film—1994	Vol 2
40E	Storage of Pyroxylin Plastic—1993	Vol 2
43B	Organic Peroxide Formulations, Storage of—1993	Vol 2
43D	Storage of Pesticides—1994	Vol 2
45	Fire Protection for Laboratories Using Chemicals—1991	Vol 2
46	Storage of Forest Products—1990	Vol 10
49	Hazardous Chemicals Data—1994	Vol 10
50	Bulk Oxygen Systems at Consumer Sites—1990	Vol 2
50A	Gaseous Hydrogen Systems at Consumer Sites—1994	Vol 2
50B	Liquefied Hydrogen Systems at Consumer Sites—1994	Vol 2
51	Design and Installation of Oxygen-Fuel Gas Systems for Welding, Cutting and Allied Processes—1992	Vol 2
51A	Acetylene Cylinder Charging Plants—1989	Vol 2
51B	Cutting and Welding Processes—1994	Vol 2
52	Compressed Natural Gas (CNG) Vehicular Fuel Systems—1995	Vol 2
53	Fire Hazards in Oxygen-Enriched Atmospheres—1994	Vol 10
54	National Fuel Gas Code—1992	Vol 2
55	Compressed and Liquefied Gases in Portable Cylinders—1993	Vol 2
58	Storage and Handling of Liquefied Petroleum Gases—1995	Vol 3
59	Storage and Handling of Liquefied Petroleum Gases at Utility Gas Plants—1995	Vol 3
59A	Liquefied Natural Gas (LNG)—1994	Vol 3
61	Fire and Dust Explosions in Agricultural and Food Products Facilities—1995	Vol 3
65	Processing and Finishing of Aluminum—1993	Vol 3
68	Venting of Deflagrations—1994	Vol 10
69	Explosion Prevention Systems—1992	Vol 3
70	National Electrical Code®—1996	Vol 3
70B	Electrical Equipment Maintenance—1994	Vol 10
70E	Electrical Safety Requirements for Employee Workplaces—1995	Vol 3
72	National Fire Alarm Code—1993	Vol 4
73	Residential Electrical Maintenance—1994	Vol 4
75	Protection of Electronic Computer/Data Processing Equipment—1995	Vol 4
77	Static Electricity—1993	Vol 10
79	Electrical Standard for Industrial Machinery—1994	Vol 4
80	Fire Doors and Fire Windows—1995	Vol 4
80A	Exterior Fire Exposures—1993	Vol 10
81	Fur Storage, Fumigation and Cleaning—1986	Vol 4
82	Incinerators, Waste and Linen Handling Systems and Equipment—1994	Vol 4
86	Ovens and Furnaces—1995	Vol 4
86C	Industrial Furnaces Using a Special Processing Atmosphere—1995	Vol 4
86D	Industrial Furnaces Using Vacuum as an Atmosphere—1995	Vol 4
88A	Parking Structures—1995	Vol 4
88B	Repair Garages—1991	Vol 4
90A	Installation of Air Conditioning and Ventilating Systems—1993	Vol 4
90B	Installation of Warm Air Heating and Air Conditioning Systems—1993	Vol 4
91	Installation of Exhaust Systems for Air Conveying of Materials—1995	Vol 4
92A	Smoke-Control Systems—1993	Vol 10
92B	Smoke Management Systems in Malls, Atria, Large Areas—1995	Vol 10
96	Ventilation Control and Fire Protection of Commercial Cooking Operations—1994	Vol 4
97	Glossary of Terms Relating to Chimneys, Vents and Heat Producing Appliances—1992	Vol 10
99	Health Care Facilities—1993	Vol 5
99B	Hypobaric Facilities—1993	Vol 5
101•	Safety to Life from Fire in Buildings and Structures—1994	Vol 5
101A	Alternative Approaches to Life Safety—1995	Vol 10
102	Grandstands, Folding and Telescopic Seating, Tents, and Membrane Structures—1995	Vol 5
105	Smoke-Control Door Assemblies—1993	Vol 10
110	Emergency and Standby Power Systems—1993	Vol 5
111	Stored Electrical Energy Emergency and Standby Power Systems—1993	Vol 5

115	Laser Fire Protection—1995	Vol 10
120	Coal Preparation Plants—1994	Vol 5
121	Self-Propelled and Mobile Surface Mining Equipment—1990	Vol 5
122	Fire Prevention and Control in Underground Metal and Nonmetal Mines—1995	Vol 5
123	Fire Prevention and Control in Underground Bituminous Coal Mines—1995	Vol 5
130	Fixed Guideway Transit Systems—1995	Vol 5
150	Fire Safety in Racetrack Stables—1995	Vol 5
170	Fire Safety Symbols—1994	Vol 5
203	Roof Coverings and Roof Deck Constructions—1995	Vol 10
204M	Smoke and Heat Venting—1991	Vol 10
211	Chimneys, Fireplaces, Vents and Solid Fuel Burning Appliances—1992	Vol 5
214	Water-Cooling Towers—1992	Vol 5
220	Types of Building Construction—1995	Vol 5
221	Fire Walls and Fire Barrier Walls—1994	Vol 5
231	General Storage—1995	Vol 5
231C	Rack Storage of Materials—1995	Vol 6
231D	Storage of Rubber Tires—1994	Vol 6
231E	Storage of Baled Cotton—1989	Vol 10
231F	Storage of Roll Paper—1987	Vol 6
232	Records, Protection of—1995	Vol 6
232A	Archives and Records Centers—1995	Vol 11
241	Construction, Alteration, and Demolition Operations—1993	Vol 6
251	Fire Tests of Building Construction and Materials—1995	Vol 6
252	Fire Tests of Door Assemblies—1995	Vol 6
253	Test for Critical Radiant Flux of Floor Covering Systems Using a Radiant Heat Energy Source—1995	Vol 6
255	Test of Surface Burning Characteristics of Building Materials—1990	Vol 6
256	Methods of Fire Tests of Roof Coverings—1993	Vol 6
257	Fire Tests of Window Assemblies—1990	Vol 6
258	Research Test Method for Determining Smoke Generation of Solid Materials—1994	Vol 6
259	Test Method for Potential Heat of Building Materials—1993	Vol 6
260	Cigarette Ignition Resistance of Components of Upholstered Furniture—1994	Vol 6
261	Method of Test for Determining Resistance of Mock-Up Upholstered Furniture Material Assemblies to Ignition by Smoldering Cigarettes—1994	Vol 6
262	Method of Test for Fire and Smoke Characteristics of Wires and Cables—1994	Vol 6
263	Heat and Visible Smoke Release Rates for Materials and Products, Method of Test for—1994	Vol 6
264	Heat and Visible Smoke Release Rates for Materials and Products Using an Oxygen Consumption Calorimeter—1995	Vol 6
264A	Method of Test for Heat Release Rates for Upholstered Furniture Components or Composites and Mattresses Using an Oxygen Consumption Calorimeter—1994	Vol 6
265	Evaluating Room Fire Growth Contribution of Textile Wall Coverings—1994	Vol 6
266	Fire Characteristics of Upholstered Furniture Exposed to Flaming Ignition Source, Method of Test for—1994	Vol 6
267	Fire Characteristics of Mattresses and Bedding Assemblies Exposed to Flaming Ignition Source, Method of Test for—1994	Vol 6
291	Fire Flow Testing and Marking of Hydrants—1995	Vol 11
295	Wildfire Control—1991	Vol 6
297	Principles and Practices for Communication Systems—1995	Vol 11
298	Foam Chemicals for Class A Fuels—Rural, Suburban, Veg. Areas—1994	Vol 6
299	Protection of Life and Property from Wildfire—1991	Vol 6
302	Pleasure and Commercial Motor Craft—1994	Vol 6
303	Marinas and Boatyards—1995	Vol 6
306	Control of Gas Hazards on Vessels—1993	Vol 6
307	Marine Terminals, Piers and Wharves—1995	Vol 6
312	Fire Protection of Vessels During Construction, Repair, and Lay-Up—1995	Vol 6
318	Protection of Cleanrooms—1995	Vol 6
321	Basic Classification of Flammable and Combustible Liquids—1991	Vol 6
325	Fire Hazard Properties of Flammable Liquids, Gases, and Volatile Solids—1994	Vol 11
326	Safe Entry of Underground Storage Tanks—1993	Vol 6
327	Cleaning or Safeguarding Small Tanks and Containers Without Entry—1993	Vol 6

Alphabetical Index—2000 National Fire Codes Contents

Notes

Preface

1. Jennifer Hemmingsen, *The Albert Lea Tribune* (Albert Lea, MN), July 9, 2001.
2. *Fire in the United States*, (Emmitsburg, MD: United States Fire Administration, National Fire Data Center, August 1999), p. 1.
3. *America at Risk*, (Emmitsburg, MD: Recommissioned Panel for America Burning, Federal Emergency Management Agency, May 2000), p. 15.
4. Ibid., p. 16.
5. National Association of Credit Men, Committee on Fire Insurance, June 1909, quoted in Harry Chase Brearley, *Fifty Years of Civilizing Force* (new York: Frederick A. Stokes, 1916), p. 111.
6. *Twenty-five Fatality Fire at Chicken Processing Plan, Hamlet, North Carolina, September 1991*, (Technical Report #057) (Emmitsburg, MD: United States Fire Administration, June 1999), p. 9.

Chapter One

1. *Field Practice—An Inspection Manual for Property Owners, Fire Departments and Inspection Offices* (Quincy, MA: National Fire Protection Association, 1922), p. XX.
2. Alexander Reid, *Aye Ready!* (Edinburgh, Scotland: George Steward and Company, Ltd., 1974), p. 5.
3. Ronny J. Coleman et al., *Managing Fire Services* (Washington, DC: International City/County Management Association, 1988), p. 8.
4. Coleman, p. 13
5. National Commission on Fire Prevention and Control, *America Burning* (Washington, DC: Government Printing Office, 1973), p. 79.
6. *BOCA Basic/National Building Code* (Country Club Hills, IL: Building Officials and Code Administrators International, 1984), p. iv.
7. Kevin Cassidy, *Fire Safety and Loss Prevention* (Stoneham, MA: Butterworth-Heinemann, 1992), p. 56.
8. BOCA: "The International Code Council," *The Building Official and Code Administrator Magazine*, March/April 1995, p. 12.
9. National Commission on Fire Prevention and Control, *America Burning*, p. 79.
10. Albert Harkness, "Building Codes: A Historical Perspective," *Building Officials and Code Administrators Magazine*, 26 (March/April 1996), p. 15.
11. Ibid., p. 79.
12. Ibid., p. 79.
13. National Fire Protection Association, *NFPA 100 Years, a Fire Protection Overview* (Quincy, MA: National Fire Protection Association, 1996), p. 9.
14. Ibid., p. 15.
15. *Code for Safety to Life from Fire in Buildings and Structures* (Quincy, MA: National Fire Protection Association, 1994), p. 17.
16. NFPA 251–95, *Standard Methods of Tests of Fire Endurance of Building Construction and Materials* (Quincy, MA: National Fire Protection Association, 1995), p. 2.
17. *NFPA Standards-Making System* (Quincy, MA: National Fire Protection Association, 1996), p. 3.

18. National Fire Academy, *Fire Prevention Organization and Management*, pp. 6–9.
19. Timothy Callahan, *Fire Service and the Law* (Quincy, MA: National Fire Protection Association, 1987), p. 134.
20. Clay L. Wirt, *Virginia Town and City*, August 1989, p. 12.
21. George Dean et al., *Legal Aspects of Code Administration* (Country Club Hills, IL: Building Officials and Code Administrators, International Congress of Building Officials, Southern Building Code Congress International, 1984), p. 16.
22. Callahan, p. 134.
23. National Fire Academy, *Fire Prevention Organization and Management*, pp. 6–13.

Chapter Two

1. "Fire Hazards," Report of the Virginia Advisory Legislative Council to the Governor and the General Assembly of Virginia, Senate Document 11 (Richmond: Commonwealth of Virginia Division of Purchase and Printing, 1948), p. 5.
2. John R. Hall et al., *Fire Code Inspections and Fire Prevention: What Methods Lead to Success?* (Boston, MA: National Fire Protection Association, 1978) p. viii.
3. George Dean, et al., *Legal Aspects of Code Administration* (Country Club Hills, IL: Building Officials and Code Administrators, International Congress of Building Officials, Southern Building Code Congress International, 1984), p. 135.
4. Timothy Callahan, *Fire Service and the Law* (Quincy, MA: National Fire Protection Association, 1987), p. 174.
5. Dean et al., p. 85.
6. Callahan, p. 178.
7. *The BOCA National Fire Prevention Code Commentary* (Country Club Hills, IL: Building Officials and Code Administrators International, 1993), p. 1–17.
8. National Fire Academy, *Political and Legal Foundations of Fire Protection* (Emmitsburg, MD: National Fire Academy, 1992), p. 9-2.
9. Howard Markman et al., *Political and Legal Foundations of Fire Prevention*, p. 9-3.
10. Ibid., p. 9-3.
11. Callahan, p. 167.
12. Dean et al., p. 80.
13. Ibid., p. 80.
14. George E. Rush, *Dictionary of Criminal Justice* (Guilford, CT: Dushkin, 1986), p. 221.
15. Dean, p. 142.
16. Rush, p. 104.
17. Ibid., p. 104.

Chapter Three

1. *Standard Building Code* (Birmingham, AL: Southern Building Code Congress International, 1994), p. 116.

Chapter Four

1. Ronny J. Coleman et al., *Managing Fire Services* (Washington, DC: International City/County Management Association, 1988), p. 7.
2. Francis L. Brannigan, *Building Construction for the Fire Service* (Boston, MA: National Fire Protection Association, 1971), p. 8.
3. *BOCA National Building Code* (Country Club Hills, IL: Building Officials and Code Administrators International, 1996), p. 65.
4. *Uniform Building Code* (Whittier, CA: International Conference of Building Officials, 1994), p. 1–17.
5. *Standard Building Code* (Birmingham, AL: Southern Building Code Congress International, 1994), p. 22.
6. *BOCA National Building Code*, p. 58.

Chapter Five

1. *Fire Inspection Principles Course Guide* (Emmitsburg, MD: National Fire Academy, 1993).
2. *Fire Protection Handbook*, 16th ed. (Quincy, MA: National Fire Protection Association), p. 6-76.
3. Ibid., p. 20.
4. *International Building Code* (Falls Church, VA: International Code Council, 2000), p. 92.

5. *BOCA National Building Code* (Country Club Hills, IL: Building Officials and Code Administrators International, 1996), p. 65.
6. Ibid., p. 65.
7. *Basic Building Code*, 5th ed. (Country Club Hills, IL: Building Officials and Code Administrators, 1970).
8. *Basic Building Code*, 5th ed. (Country Club Hills, IL: Building Officials and Code Administrators, 1981).

Chapter Six

1. *Fire Protection Handbook*, 16th ed. (Quincy, MA: National Fire Protection Association), p. 17-2.
2. *Fire Protection Handbook*, 13th ed. (Quincy, MA: National Fire Protection Association), p. 16-2.
3. *Fire Protection Handbook*, 16th ed., p. 4-42.
4. Richard E. Hughey, P. E., "Property Protection or Life Safety—Can We Have Both?", *Fire Marshal Quarterly*, September 1996, p. 20.
5. IFSTA 210, *Private Fire Protection and Detection* (Stillwater, OK: International Fire Service Training Association, 1987), p. 39.
6. *Fire Protection Handbook*, 13th ed., p. 16-40.
7. NFPA 13, *Standard for the Installation of Sprinkler Systems*, 1994 ed. (Quincy, MA: National Fire Protection Association), p. 17.
8. *BOCA National Building Code*, 1996 ed. (Country Club Hills, IL: Building Officials and Code Administrators International, 1996), p. 93.
9. Tom Multer and Richard Thieken, "Careful Planning is Necessary in ESFR Sprinkler Installations," *Engineering and HVAC Design*.
10. *High-rise Office Building Fire, One Meridian Plaza Philadelphia, Pennsylvania* (Emmitsburg, MD: United States Fire Administration), p. 15.
11. Edward J. Kaminski, P. E., "Practical Aspects of Wet and Dry Chemical Extinguishing System Inspection and Acceptance," *Building Officials and Code Administrators Magazine*, 4 (July/August 1995), p. 34.
12. R. T. Leicht, "Commercial Cooking Protection: The Myths and Facts Concerning the UL-300 Standard," *Fire Marshals Quarterly*, April 1996.
13. *Questions and Answers on Halons and Their Substitutes* (USEPA, http//www.epa.gov/ozone/title6/snap/halon_qs.html) November 2, 1996.
14. *Fire Protection Handbook*, 13th ed., p. 15-23.
15. NFPA 27, *National Fire Alarm Code*, 1996 ed. (Quincy, MA: National Fire Protection Association, 1996), p. 83.

Chapter Seven

1. *Fire Journal* (Quincy, MA: National Fire Protection Association, August 1977), p. 50.
2. Ibid., p. 50.
3. *Uniform Building Code* (Whittier, CA: International Congress of Building Officials, 1994), p. 1-173.
4. *Standard Building Code* (Birmingham, AL: Southern Building Code Congress International, 1994), p. 23.
5. *Security Bars and Fire Safety* (Quincy, MA National Fire Protection Association), 1996.

Chapter Eight

1. "Flammable Decorations, Lack of Exits Create Tragedy at Coconut Grove," *Fire Engineering*, August, 1977, p. 67.
2. Ibid., p. 72.
3. *Fire Inspection and Code Enforcement* (Stillwater, OK: Fire Protection Publications, 1987), p. 75.
4. David P. Demars, *Fire Investigation Report, Hotel Fire, Las Vegas Nevada, February 10, 1981* (Quincy, MA: National Fire Protection Association, 1981).

SECTION THREE: THE FIRE PREVENTION CODE

1. *Uniform Fire Code*, 1994 ed. (Whittier, CA: International Fire Code Institute, 1994), p. 1-1.

Chapter Nine

1. NFPA 550, *Fire Safety Concepts Tree* (Quincy, MA: National Fire Protection Association, 1986), p. 14.
2. *Fire Safe Building Design* (Emmitsburg, MD: National Fire Academy, 1996).
3. *Dibble v. Brunty* United States District Court for the

Eastern District of Virginia, Civil Action No. 95-356-A.

4. *Fire Prevention Code Commentary* (Country Club Hills, IL: Building Officials and Code Administrators, 1993), p. 3–5.
5. Alisa Wolf, "Fraternity Fire Kills Five," *NFPA Journal*, September/October 1996, p. 61.
6. Ibid., p. 61.
7. *Fire in the United States* (Emmitsburg, MD: National Fire Data Center, 1983–1990), p. 231.
8. *Uniform Fire Code*, 1994 ed. (Whittier, CA: International Fire Code Institute, 1994), p. 1-46, 902.2.1.
9. *BOCA National Building Code*, 1996 ed. (Country Club Hills, IL: Building Officials and Code Administrators, 1996), p. 57, 506.2.
10. *BOCA National Fire Prevention Code*, 1996 ed. (Country Club Hills, IL: Building Officials and Code Administrators International, 1996), p. 18, F-3111.1.

Chapter Ten

1. NFPA 25, *Standard for the Inspection, Testing and Maintenance of Water-Based Fire Protection Systems* (Quincy, MA: National Fire Protection Association, 1995), p. 47.

Chapter Twelve

1. NFPA 704, *Identification of the Fire Hazards of Materials* (Quincy, MA: National Fire Protection Association, 1990).

Chapter Thirteen

1. *Compton's Reference Collection*, Compton's New Media, Inc., 1995.
2. *Fire Prevention Code Commentary* (Country Club Hills, IL: Building Officials and Code Administrators International, 1993), p. 32–1.
3. John D. DeHaan, *Kirk's Fire Investigation* (Englewood Cliffs, NJ: Prentice-Hall, 1991), p. 44.
4. Ibid., p. 36.
5. Richard Best, "$100 Million Fire in K Mart Distribution Center," *Fire Journal*, March 1983, p. 36.
6. NFPA 30B, *Code for the Manufacture and Storage of Aerosol Products* (Quincy, MA: National Fire Protection Association, 1994), p. 24.

Chapter Fourteen

1. Robert C. Morehard, *Explosives and Rock Blasting* (Dallas, TX: Atlas Powder Company, 1987), p. 1.
2. James H. Meidl, *Explosive and Toxic Hazardous Materials* (Beverly Hills, CA: Glencoe Press, 1972), p. 51.
3. Morehard, p. 3.
4. *Compton's Interactive Encyclopedia* (Compton's New Media, Inc., 1995).
5. Morehard, p. 330.
6. NFPA 495, *Explosive Materials Code* (Quincy, MA: National Fire Protection Association, 1996), p. 23.
7. Morehard, p. 5.
8. Meidl, p. 4.
9. Morehard, p. 300.
10. Alert Bulletin, *Multiple-death Fire Fighter Fatality Incident Kansas City, Missouri*, National Fire Protection Association, December 7, 1988.
11. Ibid.
12. *The BOCA National Fire Prevention Code Commentary*, (Country Club Hills, IL: Building Officials and Code Administrators International, 1993), p. 31-1.

Chapter Fifteen

1. Alisa Wolf, "Seventeen Die in Dusseldorf Airport Terminal Fire," *NFPA Journal*, July/August 1996.
2. *The BOCA National Fire Prevention Code Commentary*, (Country Club Hills, IL: Building Officials and Code Administrators International, 1993), p. 8-1.

Chapter Sixteen

1. Stephen G. Badger, "1995 Large-Loss Fires and Explosions," *NFPA Journal*, November/December 1996.
2. *Before the Fire—Fire Prevention Strategies for Storage Occupancies* (Quincy, MA: National Fire Protection Association, 1988), p. 8.

3. Jack Yates, "Chicken Processing Plant Fires," U.S. Fire Administration, Report #57.
4. Ibid.
5. James L. Linville et al., *Industrial Fire Hazards Handbook* (Quincy, MA: National Fire Protection Association, 1984), p. 974.
6. Ibid., p. 986

Chapter Seventeen

1. James H. Meidl, *Flammable Hazardous Materials* (Beverly Hills, CA: Glencoe Press, 1970), p. 93.
2. Compressed Gas Association, *Handbook of Compressed Gases* (New York: Van Nostrand Reinhold, 1981), p. 5.
3. John D. DeHaan, *Kirk's Fire Investigation* (Englewood Cliffs, NJ: Prentice-Hall, 1991), p. 371.
4. *NFPA Fire Protection Handbook*, 18th ed. (Quincy, MA: National Fire Protection Association), pp. 4–72.
5. Ibid.
6. Meidl, p. 96.
7. Ibid., p. 45.
8. Ibid., p. 134.
9. Ibid., p. 146.

Chapter Eighteen

1. George W. Rambo, "Pest Control," *New York Public Library Desk Reference*, Compton's New World, 1996.
2. Richard Stilp and Armando Bevelacqua, *Emergency Medical Response to Hazardous Materials Incidents* (Albany: Delmar Publishers, 1997), p. 274.

Glossary

Accelerator A quick opening device that permits system air pressure to enter the dry pipe valve below the one-way clapper valve, unbalancing the differential and causing the valve to trip more quickly.

Accessible means of egress A means of egress including the exit access, exit, and exit discharge that can be entered and used by a person with a severe disability using a wheelchair and is also safe and usable for people with other disabilities.

Accessory use area A portion of a building with a different use group classification than the main area, but that does not require a fire separation between the main areea, due to its small size.

Adopt To formally accept and put into effect.

Adoption by reference Method of code adoption in which the specific edition of a model code is referred to within the adopting ordinance.

Adoption by transcription Method of code adoption in which the entire text of the code is published within the adopting ordinance.

Aerosol A product that is dispensed from a container by means of a liquified or compressed gas.

Ambulatory Able to walk about without assistance. Capable of sensing an emergency situation and appropriately responding by exiting the building.

Approved Acceptable to the code official with jurisdiction. Approval is normally based on nationally recognized standards or, in their absence, on sound engineering practice.

Assembly (rated assembly) A building component such as a door, wall, damper, or ceiling composed of specific parts and tested and listed as a unit.

AST Aboveground storage tank for regulated liquids.

Blasting agents Materials or mixtures containing fuel and an oxidizer not otherwise classified as an explosive and that cannot be detonated by means of a #8 blasting cap when unconfined.

Board foot A measurement used for lumber equalling 144 cubic inches (12 inches by 12 inches by 1 inch).

Building height According to the *IBC*, "the vertical distance from grade plane to the average height of the highest roof surface."

Citation A written order, issued by a law enforcement officer or other authorized official, directing an alleged offender to appear in court at a specific time to answer a criminal charge.

Closure Tank closure; placing a tank permanently out of service by removal or abandonment in place using an approved method.

Code A systematically arranged body of rules. When and where to do or not to do something.

Combustible liquid A liquid having a flash point at or above 100°F. Combustible liquids are further categorized based on flash point as types II, IIIA, and IIIB.

Common path of travel The portion of an exit access that building occupants must traverse before two distinct paths of travel or two exits are available.

Compressed gas A gas or mixture of gases having an absolute pressure exceeding 40 psi at 70°F, or an absolute pressure exceeding 140 psi at 130°F, or any liquid with a vapor pressure that exceeds 40 psi at 100°F.

Container A vessel with a capacity of 60 gallons or less used for the storage or transportation of flammable or combustible liquids. Piping and engine fuel tanks are not containers.

Control areas Areas within a building in which hazardous materials in quantities not exceeding the exempt amounts may be stored, handled, or used.

Cryogenic liquid A refrigerated liquid gas with a boiling point below –130°F (SFPC 94) or below –150°F (UPC 94) or below –200°F (BNFPC 96). Also known as cryogenic fluid.

Dead end A corridor, hallway, or passageway open to a corridor that can be entered from the exit access without passage through a door, but that does not lead to an exit.

Deflagration Very rapid but subsonic oxidation evolving heat, light, and a low-energy pressure wave that is capable of causing damage.

Deluge sprinkler system An automatic sprinkler system that features open heads, dry piping, and a deluge valve that controls the supply of water, designed to wet the entire area upon activation.

Deluge valve Sprinkler water supply valve that is activated automatically or manually.

Density Refers to sprinkler density calculated by gallons per minute discharge divided by the square footage covered.

Design occupant load The number of persons expected to occupy a space based on design tables in the building code. The number and capacity of exits is based on the design occupant load.

Detection device A device connected to a fire alarm system having a sensor that responds to physical stimulus such as heat or smoke.

Detonator A device, consisting of electric and nonelectric blasting caps, fuse caps, and detonating cord delay connectors, that contains a primary or initiating explosive designed to set off an explosive reaction.

Detonator, delay type A detonator that introduces a specific time lapse between the firing signal and detonation of the main explosive charge.

Detonator, instantaneous type A detonator with no time lapse between the firing signal and the detonation of the main explosive charge.

Differential Ratio of air pressure to water pressure that is necessary to balance a dry pipe valve, maintaining it in the closed position.

Dillon's rule Legal ruling issued by Chief Justice John Forrest Dillon of the Iowa Supreme Court in the late 1800s whereby local governments possess only those powers expressly granted by charter or statute.

Dry pipe sprinkler system An automatic sprinkler system that features dry piping maintained under constant air pressure and a dry pipe valve in which water is held back by the pressure differential between the system air pressure and water supply pressure. Used where there is a danger of freezing.

Dry pipe valve Sprinkler water supply valve designed to permit a moderate amount of air pressure above the valve to hold back a much greater water pressure from the incoming supply.

Early suppression fast response sprinklers (ESFR) A type of fast response sprinkler designed to suppress fires in high challenge fire hazards through the application of increased flow densities.

Exempt amount Threshold quantity of a hazardous material established by the building and fire prevention codes as the maximum amount that can be stored, handled, or used within a building that is not classified as Use Group H.

Exhauster A quick opening device used in dry pipe sprinkler systems that uses an auxiliary valve to discharge system air pressure to the atmosphere.

Exit That portion of a means of egress that is separated from all other parts of a building by rated assemblies and provides a protected path to the exit discharge.

Exit access That portion of a means of egress from any point in a building to an exit.

Exit discharge That portion of a means of egress between the exit and public way.

Explosive materials Explosives, blasting agents, and detonators, including dynamites, slurries, emulsions and water gels, black powder, smokeless powder, detonators and safety fuses, squibs, detonating cord, and other materials whose primary function is to function by explosion.

Explosives Chemical compounds or mixtures whose primary function is to function by explosion and that cause a sudden and almost instantaneous release of pressure, gas, and heat.

Fire Rapid oxidation accompanied by the evolution of heat and light.

Fire line Dedicated underground supply piping for a sprinkler or standpipe system.

Fire protection system Equipment and devices designed to detect a fire, sound an alarm and (or) make notification, control or remove smoke and hot gases, and control or extinguish the fire.

Fire resistance The resistance of a building to collapse or to total involvement in fire; the property of materials and their assemblies that prevents or retards the passage of excessive heat, hot gases, or flames under conditions of use.

Flame resistant The resistance of a material to ignition and combustion when exposed to a small ignition source. Some materials are inherently flame resistant, while others are treated through impregnating or coating.

Flammable liquid A liquid having a flash point below 100°F and further categorized based on flash point and boiling point as types IA, IB, and IC.

Flashover Point at which the contents of a room or space becomes heated, simultaneously ignites, and the entire room or space becomes involved in fire.

Flash point The minimum temperature at which a liquid gives off sufficient vapor to form an ignitable mixture at the surface, but not sufficient to sustain combustion.

Floor area Calculation used in the *UBC* in determining the design occupant load. Floor area includes the area within the exterior walls exclusive of vent shafts, courts, and accessory spaces ordinarily used only by occupants of the main area.

Flyrock Rock propelled from the blast area by blasting operations.

Freedom of Information Act A law passed by Congress and signed by President Lyndon Johnson in 1967. The law, and state laws patterned after it, guarantee public access to all documents and information under control of the government, with certain specific exemptions.

Grade The average of finished ground level around a building or within a given distance from a building. Grade is used in determining building height.

Grade plane An imaginary plane repersenting the finished ground around the exterior walls of a building.

Gross floor area Calculation used by the *BNBC* and *SBC* in determining the design occupant load of spaces with low occupancy densities, such as business, industrial, and mercantile areas. Gross floor area includes the entire area within the exterior walls.

Halogenated extinguishing agent (halon) A clean extinguishing agent composed of carbon and one or more elements from the halogen series (fluorine, chlorine, bromine, and iodine), which leaves no residue.

Hazardous fire area Term used in the *UFC* to describe public and privately owned areas of grass, brush, or forest with limited accessibility such that a fire originating in the area would be unusually difficult to extinguish and lead to significant damage and potential erosion.

Hazardous Materials Inventory Statement (HMIS) An inventory of regulated materials in a form approved by the fire official that includes the following information: chemical name; manufacturer's name; hazardous ingredients; UN, NA, or CIS identification number; and maximum quantities stored or used on site at any time.

Hazardous Materials Management Plan (HMMP) A management and contingency plan in a form approved by the fire official that includes the following information: site plan, floor plan, material compatibility, monitoring and security methods, hazard identification, inspection procedures, employee training, and emergency equipment available.

Hazardous production material (HPM) A solid, liquid, or gas that is classified as a 3 or 4 in accordance with NFPA 704 for hazards to health, flammability, or reactivity and used in research or production where the end product is not hazardous.

Health hazard material A material that affects the target organs of the body including the liver, kidneys, central nervous system, reproductive system, and circulatory system.

Heat of combustion The amount of heat given off by a particular substance during the combustion process. A measure of fuel efficiency.

Heliport An area where helicopters take off and land from the ground or water, or from a building. Includes areas for storage, maintenance, and refueling.

Helistop Same as a heliport, except without the facilities for storage, maintenance, or refueling.

Horizontal exit An exit from one building to another on approximately the same level; or a passage through or around a rated wall or partition that affords protection from fire or smoke coming from the area from which escape is made.

Hydrant coefficient A number describing the opening from which water flows from a fire hydrant.

Injunction A legal order issued by a court that commands a person or entity to perform a specific act or prohibits a specific action by that person or entity.

Irritant A noncorrosive chemical that has a reversible inflammatory effect on living tissue at the point of skin contact.

Label A permanent identification affixed to a product by a manufacturer indicating the function and performance characteristics of the product, name of the manufacturer, and the name of the approved testing agency that tested a representative sample of the product.

Limited area sprinkler system An automatic sprinkler system that is limited to a single fire area and consists of not more than twenty sprinklers.

Listed Equipment or materials included in a document prepared by an approved testing agency indicating that the equipment or materials were tested in accordance with an approved test protocol and found suitable for a specific use.

Magazine A structure designed and constructed for the storage of explosive materials.

Mandatory Code provisions that must be enforced and complied with.

Material Safety Data Sheet (MSDS) A document, prepared in accordance with DOL 29 CFR, that contains information regarding the physical and health hazards associated with a given product or substance and a recommended emergency action.

May Indicates a discretionary provision of the code. Enforcement is left to the judgment of the fire official.

Maximum allowable quantity See Exempt amount.

Model code A code developed by an organization for adoption by governments. The Uniform, Standard, and BOCA/National codes are examples. Model codes are generally developed through the consensus process through the use of technical committees.

Mini/maxi code A code developed and adopted at the state level for either mandatory or optional enforcement by local governments, and one that cannot be amended by the local governments.

Negligence Culpable carelessness; the failure to act as a reasonable and prudent person under similar circumstances.

Nesting Method of securing compressed gas cylinders in groups, in which each cylinder has a minimum of three points of contact with other cylinders, walls, or bracing.

Net floor area Calculation used by the *BNBC* and *SBC* in determining the design occupant load of spaces with high occupancy densities, such as assembly and educational areas. Net floor area includes the area within the walls exclusive of the thickness of walls or columns or accessory spaces such as stairs, rest rooms, or mechanical rooms.

Nonsegregated storage The storage of aerosol products in general purpose warehouses within areas that are not used exclusively for the storage of aerosols.

Opening protective A rated assembly such as a door or window that provides a protected opening in a rated wall or partition. There are interior as well as exterior opening protectives.

Ordinance A law of an authorized subdivision of a state, such as a county, city, or town.

Overcrowding A condition in which the number of occupants within a building or space exceeds the approved occupant load calculated in accordance with the building code.

Performance code A code that assigns an objective to be met and establishes criteria for determining compliance. Examples are requirements for fire assemblies rated in "hours" and building materials rated as "noncombustible" when tested to specific protocols.

Permissive Code language that is discretionary and leaves application to the judgment of the enforcing authority.

Pitot pressure The pressure (in psi) shown on the bourdon gauge attached to the pitot tube inserted in a water stream flowing from a fire hydrant.

Preaction sprinkler system An automatic sprinkler system that features dry piping, standard fusible heads, and a water supply control valve that is activated by fire detection devices. Used to reduce the possibility of water damage from accidental breakage or discharge.

Primers Explosives packages made up of an explosive charge and a detonator or detonating cord used to initiate other less sensitive explosives or blasting agents.

Protected Shielded from the effects of fire by encasement. Concrete, gypsum, and sprayed-on fire resistive coatings are all used to "protect" structural elements.

Reference datum A plane representing the elevation of the highest adjoining sidewalk or ground surface within a 5-foot horizontal distance of a building. Used by the *UBC* in determining building height.

Residual perssure The pressure (in psi) measured at the test hydrant with water flowing form the flow hydrant.

Scaled distance formula A formula used to determine maximum amount of explosive material that can be detonated per delay interval of 8 milliseconds or greater, based on distance to the nearest occupied structure.

Segregated storage The storage of aerosol products within areas specifically designed for that use incorporating code-required storage arrangements and fire protection features.

Sensitizer A chemical that causes many humans or animals to develop allergic reactions after repeated contact.

Shall Indicates a positive and definitive requirement of the code that must be performed. Action is mandatory.

Signaling device Notification appliance. An alarm system component such as a bell, horn, speaker, light, or text display that provides audible, visible, or tactile output.

Smokeproof or pressurized enclosure An enclosed exit stair connected to all floors by either exterior balconies or ventilated vestibules and designed to limit the movement of smoke and fire gases into the stairwell.

Specification code a code that specifies a type of construction or materials to be used.

Spray area An area designed and constructed to be used for the application of flammable finishes complying with the requirements of the building code.

Spray booth A structure designed and constructed to be used for the application of flammable finishes, featuring power ventilation, fixed fire suppression, and separation that is installed within a building.

Spray room A room designed and constructed for the application of flammable finishes complying with the requirements of the building code.

Standard A rule for measuring or a model to be followed. How to do something, what materials to use. Also known as *referenced standard.*

Standard time temperature curve (STTC) Curve representing the standard reproducible test fire used since 1918 to measure the fire resistance of building materials.

Standpipe pressure-regulating device A valve permanently attached to the standpipe discharge and designed to reduce flow pressure to a predetermined level by restricting the orifice size.

Static pressure Pressure exerted within a water system at no flow.

Statute A law enacted by a state or the federal legislature.

Steiner tunnel test Test to determine the surface burning characteristics of building materials in which the flame spread of the test material is compared to asbestos cement board, rated 0, and red oak, rated 100. The higher the rating, the greater the potential hazard. In addition to flame spread, the test also measures smoke development and fuel contributed to the fire. ASTM E84, NFPA 255, UL 723,

Stemming Inert material placed in the borehole after the explosive material has been loaded to confine the effects of the explosive reaction. May also separate charges within a single borehole.

Summons A written order, issued by a judicial officer, law enforcement officer, or other authorized official, directing an alleged offender to appear in court at a specific time to answer a criminal charge. A summons issued by a law or code enforcement officer is classified as a *citation.*

Thermal lag The difference between the operating temperature of a fire detection device such as a sprinkler head and the actual air temperature when the device activates.

Travel distance The length of the path a building occupant must travel before reaching an exterior door or an enclosed exit stairway, exit passageway, or horizontal exit. The total length of the exit access.

Use group Building code classification system whereby buildings and structures are grouped together by use and by the characteristics of their occupants.

UST Underground storage tank for regulated liquids.

Vapor density The ratio of the weight of a given volume of a gas to that of air.

Warrant A legal writ issued by a judicial officer commanding an officer to arrest a person, seize property, or search a premises.

Water-miscible Water soluble.

Wet pipe sprinkler system An automatic sprinkler system in which the supply valves are open and the system is charged with water under supply pressure at all times.

Acronyms

AIA	American Insurance Association
ANFO	ammonium nitrate and fuel oil
ANSI	American National Standards Institute
AST	aboveground storage tank
ASTM	American Society for Testing and Materials
BNBC	BOCA National Building Code
BNFPC	BOCA National Fire Prevention Code
BOCA	Building Officials and Code Administrators International
CABO	Council of American Building Officials
DOT	Department of Transportation
ESFR	early suppression fast response
FIFRA	Federal Insecticide, Fungicide and Rodenticide Act
HMIS	Hazardous Materials Inventory Statement
HMMP	Hazardous Material Management Plan
HPM	hazardous production material
HRC	Halon Recycling Corporation
IBC	International Building Code
ICBO	International Conference of Building Officials
IFC	International Fire Code
ICC	International Code Council
IFCI	International Fire Code Institute
IME	Institute for Makers of Explosives
LPG	liquified petroleum gas
MSDS	Material Safety Data Sheets
NAFTA	North American Free Trade Agreement
NBFU	National Board of Fire Underwriters
NEC	National Electrical Code
NFPA	National Fire Protection Association
NG	natural gas
SBC	Standard Building Code
SBCCI	Southern Building Code Congress International
SCBA	self-contained breathing apparatus
SFPC	Standard Fire Prevention Code
STTC	standard time temperature curve
UBC	Uniform Building Code
UFC	Uniform Fire Code
UL	Underwriters' Laboratories
UST	underground storage tank

Index